WIRELESS COMMUNICATIONS DESIGN HANDBOOK

Aspects of Noise, Interference, and Environmental Concerns

VOLUME 2: TERRESTRIAL AND MOBILE INTERFERENCE

WIRELESS COMMUNICATIONS DESIGN HANDBOOK
Aspects of Noise, Interference, and Environmental Concerns

VOLUME 2: TERRESTRIAL AND MOBILE INTERFERENCE

REINALDO PEREZ
Spacecraft Design
Jet Propulsion Laboratory
California Institute of Technology

ACADEMIC PRESS
San Diego   London   Boston
New York   Sydney   Tokyo   Toronto

Copyright © 1998 by Academic Press

ACADEMIC PRESS
525 B Street, Suite 1900, San Diego, CA 92101-4495, USA
http://www.apnet.com

Academic Press
24–28 Oval Road, London NW1 7DX, UK
http://www.hbuk.co.uk/ap/

Library of Congress Cataloging-in-Publication Data

Perez, Reinaldo.
Wireless communications design handbook : aspects of noise, interference, and environmental concerns / Reinaldo Perez.
p. cm.
Contents: v. 1. Space interference — v. 2. Terrestrial and mobile interference — v. 3. Interference into circuits.
ISBN 0-12-550721-6 (volume 1); 0-12-550723-2 (volume 2); 0-12-550722-4 (volume 3)
1. Electromagnetic interference. 2. Wireless communication systems—Equipment and supplies. I. Title.
TK7867.2.P47   1998   98-16901
621.382'24–dc21   CIP

Printed in the United States of America
98 99 00 01 02 IP 9 8 7 6 5 4 3 2 1

Contents

Acknowledgments

I would like to acknowledge the cooperation and support of Dr. Zvi Ruder, Editor of Physical Sciences for Academic Press. Dr. Ruder originally conceived the idea of having a series of three volumes to properly address the subject of noise and interference concerns in wireless communications systems.

Considerable appreciation is extended to Madeline Reilly-Perez who spent many hours typing, organizing, and reviewing this book.

I also want to extend my gratitude and acknowledgment to Professor Jacques Gavon of the Center for Technological Education Holon, Israel, for his help and presentation of some material in this book that comes from his own work.

Preface

Interference issues in wireless communications will also be present at the system level and are another important consideration in the main goal of addressing all aspects of noise interference and noise concerns in satellites and electronic assemblies. In this volume we cover aspects of interference that are (1) of propagation in nature, (2) are related to mobile systems, and (3) are related to the space environment.

One important aspect of interference in communications systems deals with obtaining not only the right signal, but one with sufficient strength for the customer. This scenario brings a considerable number of issues about interference for base (or ground) stations and between ground stations and mobile units. In addition, antenna design plays an important role in these interference problems, and some of these designs are discussed herein. The first part of the book addresses base stations, RF communications systems, and antenna interference problems. Mutual interference between terrestrial stations above 200 MHz, and especially above 1 GHz (for PCS and cellular), is discussed in conjunction with interference between broadcasting stations or satellite earth stations. General interference scenarios and adaptive interference cancellers are also examined.

Attention is shifted in the book to the study of popular wireless personal communication devices such as pagers and cellular phones. We not only discuss the architecture of such devices in fairly good detail (for cellular phones we discuss analog and digital phone architecture); in addition, we address proper methods for modeling antennas and matching networks for pagers and cellular phones. A certain amount of effort is also spent on GSM, PDC, and IS-54 TDMA architectures. From PCS devices we extend to the study of base-station antenna performance. Among the topics discussed are basic principles and configurations, suitable antennas for base stations, and interference scenarios among antennas, including antenna selection environmental factors.

A considerable amount of space is devoted to the study of interference scenarios and terminology such as diffraction scattering, path loss, reflection of paths, and multipath interference. Minimization of losses in open environments and enclosed building structures is addressed. Propagation models for simulating interference are then discussed in more detail. Such models include fading basics, multipath interference, Doppler spread basics, time delay spread basics, path loss, co-channel and adjacent channel interference, and a good overview of Rayleigh fading.

Analytical and computational techniques are discussed, including the application of diffraction theory and geometrical optics for modeling propagation in urban environments. Terminology such as specular reflection and reflection coefficients are first introduced. Geometrical optics, the geometrical theory of diffraction, and the uniform theory of diffraction are addressed, especially when studying the diffraction of building corners. Propagation effects on interference are also discussed. Potential interference between earth stations is considered, including the two main propagation modes, one involving propagation over near-great-circle paths and one involving scattering from rain. Among the subjects discussed in these propagation models are signal-to-interference ratio, coordination area based on great-circle propagation, coordination area for scattering by rain, interference between space and surface stations, and procedures for interference analysis. Finally, the role of propagation phenomena in earth–space telecommunications system design is illustrated, including a look at the bit-error rate of digital and analog systems, allocation of noise and signal-to-noise ratio, diversity reception, and a detailed look at telecommunications link budgets.

Two additional major topics are covered in this volume. The first topic deals with space environmental effects in communications. Some generalities are discussed, such as the natural radio noise environment and the concepts of noise factor, antenna temperature, brightness temperature, and thermal radiation. More detailed concepts of atmospheric noise are then discussed, including extraterrestrial noise. The plasmasphere as the source of space storms and the effects of such storms in satellites are addressed. The second topic addresses electromagnetic interference and receiver modeling. The approach followed is at the system level following the content of total energy. Nonlinear interference models are then explained in great detail, starting from the use of Volterra series and extending to sinusoidal steady-state and two-tone sinusoidal response within a weakly nonlinear system. Other nonlinear interference models are also covered, such as desensitization, intermodulation (second-order, third-order, fifth-order), cross-modulation, and spurious responses.

Some fundamental principles of radio systems and environmental effects are also addressed in this volume, including a basic description of contamination sources in radio systems, followed by noise-source effects and analysis, the characteristics of the radio propagation medium, and the relationship of such propagation with antenna characteristics. An imported collorary to radio system fundamentals is the study of radio system parameters and performance criteria for computation and mitigation of electromagnetic interference effects. Included in such work is a review of modulation methods and their effect on interference for both transmitters and receivers. Computational and mitigation techniques are also discussed.

From the hardware point of view and relating most to base or earth stations, this volume dedicates considerable effort to the proper grounding, bonding, and shielding of ground-station facilities. The grounding includes grounding of equipment and high-frequency phenomena. The importance of bonding path impedence is also made relevant to this work. The concepts of grounding loops and coupled ground loops are explained, together with earthing of equipment and facilities, and signal grounding. Real grounding configurations are addressed, such as daisy-chain grounding, single-point grounding, tree grounding, multipath grounding, hybrid grounding, and structural grounding.

We hope this volume will provide wireless communications engineers with a broad introduction to system-level interference problems in ground mobile units.

Introduction

The information age, which began its major drive at the beginning of the 1980s with the birth of desktop computing, continues to manifest itself in many ways and presently dominates all aspects of modern technological advances. Personal wireless communication services can be considered a "subset technology" of the information age, but they have also gained importance and visibility over the past 10 years, especially since the beginning of the 1990s. It is predicted that future technological advancements in the information age will be unprecedented, and a similar optimistic view is held for wireless personal communications. Over the past few years (since 1994), billions of dollars have been invested all over the world by well-known, technology-driven companies to create the necessary infrastructure for the advancement of wireless technology.

As the thrust into wireless personal communications continues with more advanced and compact technologies, the risks increase of "corrupting" the information provided by such communication services because of various interference scenarios. Although transmission of information through computer networks (LAN, WAN) or through wires (cable, phone, telecommunications) can be affected by interference, many steps could be taken to minimize such problems, since the methods of transmitting the information can be technologically managed. However, in wireless communications, the medium for transmission (free space) is uncontrolled and unpredictable. Interference and other noise problems are not only more prevalent, but much more difficult to solve. Therefore, in parallel with the need to advance wireless communication technology, there is also a great need to decrease, as much as possible, all interference modes that could corrupt the information provided.

In this handbook series of three volumes, we cover introductory and advanced concepts in interference analysis and mitigation for wireless personal communications. The objective of this series is to provide fundamental knowledge to system and circuit designers about a variety of interference issues which could pose potentially detrimental and often catastrophic threats to wireless designs. The material presented in these three volumes contains a mixture of basic interference fundamentals, but also extends to more advanced topics. Our goal is to be as comprehensive as possible. Therefore, many various topics are covered. A systematic approach to studying and understanding the material presented should provide the reader with excellent technical capabilities for the design, development, and manufacture of wireless communication hardware that is highly immune to interference problems and capable of providing optimal performance.

The present and future technologies for wireless personal communications are being demonstrated in three essential physical arenas: more efficient satellites, more versatile fixed ground and mobile hardware, and better and more compact electronics. There is a need to understand, analyze, and provide corrective measures for the kinds of interference and noise problems encountered in each of these three technology areas. In this handbook series we provide comprehensive knowledge about each of three technological subjects. The three-volume series, *Wireless Communications Design Handbook: Aspects of Noise, Interference, and Environmental Concerns,* includes *Volume I, Space Interference; Volume II, Terrestrial and Mobile Interference;* and *Volume III, Interference into Circuits.* We now provide in this introduction a more detailed description of the topics to be addressed in this handbook.

Volume 1

In the next few years, starting in late 1997, and probably extending well into the next century, hundreds of smaller, cheaper (faster design cycle), and more sophisticated satellites will be put into orbit. Minimizing interference and noise problems within such satellites is a high priority. In Volume 1 we address satellite system and subsystems-level design issues which are useful to those engineers and managers of aerospace companies around the world who are in the business of designing and building satellites for wireless personal communications. This material could also be useful to manufacturers of other wireless assemblies who want to understand the basic design issues for satellites within which their hardware must interface.

The first volume starts with a generalized description of launch vehicles and the reshaping of the space business in general in this post-Cold War era. A description is provided of several satellite systems being built presently for worldwide access to personal communication services. Iridium, Globalstar, Teledesic, and Odessey systems are described in some detail, as well as the concepts of LEO, MEO, and GEO orbits used by such satellite systems.

Attention is then focused briefly on the subject of astrodynamics and satellite orbital mechanics, with the sole objective of providing readers with some background on the importance of satellite attitude control and the need to have a noise-free environment for such subsystems. Volume 1 shifts to the study of each spacecraft subsystem and the analysis of interference concerns, as well as noise mitigation issues for each of the satellite subsystems. The satellite subsystems addressed in detail include attitude and control, command and data handling, power (including batteries and solar arrays), and communications. For each of these subsystems, major hardware assemblies are discussed in detail with respect to their basic functionalities, major electrical components, typical interference problems, interference analysis and possible solutions, and worst-case circuit analysis to mitigate design and noise concerns.

Considerable attention is paid to communications subsystems: noise and interference issues are discussed for most assemblies such as transponders, amplifiers, and antennas. Noise issues are addressed for several multiple access techniques used in satellites, such as TDMA and CDMA. As for antennas, some fundamentals of antenna theory are first addressed with the objective of extending this work to antenna interference coupling. The interactions of such antennas with natural radio noise are also covered. The next subject is mutual interference phenomena affecting space-borne receivers. This also includes solar effects of VHF communications between synchronous satellite relays and earth ground stations. Finally, satellite antenna systems are discussed in some detail.

The final section of Volume 1 is dedicated to the effects of the space environment on satellite communications. The subject is divided into three parts. First, the space environment, which all satellites must survive, is discussed, along with its effects on uplink and downlink transmissions. Second, charging phenomena in spacecraft are discussed, as well as how charging could affect the noise immunity of many spacecraft electronics. Finally, discharging events are investigated, with the noise and interference they induce, which could affect not only spacecraft electronics, but also direct transmission of satellite data.

Volume 2

In the second volume of this handboook series, attention is focused on system-level noise and interference problems in ground fixed and mobile systems, as well as personal communication devices (e.g., pagers, cellular phones, two-way radios). The work starts by looking at base station RF communications systems and mutual antenna interference. Within this realm we address interference between satellite and earth station links, as well as interference between broadcasting terrestrial stations and satellite earth stations. In this approach we follow the previous work with a brief introduction to interference canceling techniques at the system level.

Volume 2 devotes considerable space to base-station antenna performance. We address, in reasonably good technical detail, the most suitable antennas for base-station design and how to analyze possible mutual interference coupling problems. The book also gives an overview of passive repeater technology for personal communication services and the use of smart antennas in such systems.

A section of Volume 2 is dedicated entirely to pagers and cellular phones and interference mitigation methods. The fundamentals of pagers and cellular phone designs are studied, and the use of diversity in antenna design to minimize interference problems is reviewed.

A major section of this volume starts with the coverage of propagation models for simulating interference. In this respect we cover Rayleigh fading as it relates to multipath interference. Path loss, co-channel, and adjacent channel interference follows. This last material is covered in good detail, since these techniques are prevalent in the propagation models used today.

The last sections of Volume 2 deal in depth with the subject of path loss, material that needs better coverage than found in previous books. The following subjects are reviewed in detail: ionospheric effects, including ionospheric scintillation and absorption; tropospheric clear-air effects (including refraction, fading, and ducting); absorption, scattering, and cross-polarization caused by precipitation; and an overall look at propagation effects on interference.

Volume 3

In Volume 3, we focus our attention inward to address interference and noise problems within the electronics of most wireless communications devices. This is an important approach, because if we can mitigate interference problems at some of the fundamental levels of design, we could probably take great steps toward diminishing even more complex noise problems at the subsystem and system levels. There are many subjects that could be covered in Volume 3. However, the material that has been selected for instruction is at a fundamental level and useful for wireless electronic designers committed to implementing good noise control techniques. The material covered in Volume 3 can be divided into two major subjects: noise and interference concerns in digital electronics, including mitigation responses; and noise and interference in analog electronics, as well as mitigation responses. In this volume we also address computational electromagnetic methods that could be used in the analysis of interference problems.

In the domain of digital electronics we devote considerable attention to power bus routing and proper grounding of components in printed circuit boards (PCBs). A good deal of effort is spent in the proper design of power buses and grounding configurations in PCBs including proper layout of printed circuit board traces, power/ground planes, and line impedance matches. Grounding analysis is also extended to the electronic box level and subsystem level, with the material explained in detail. At the IC level we concentrate in the proper design of ASIC and FPGA to safeguard signal integrity and avoid noise problems such as ground bounce and impedance reflections. Within the area of electronic design automation (EDA), parasitics and verification algorithms for ASIC design are also discussed. A great deal of effort is put into the study of mitigation techniques for interference from electromagnetic field coupling and near-field coupling, also known as crosstalk, including crosstalk among PCB card pins of connectors. The work continues with specific analysis of the interactions in high-speed digital circuits concerning signal integrity and crosstalk in the time domain. Proper design of digital grounds and the usage of proper bypass capacitance layout are also addressed. Other general topics such as power dissipation and thermal control in digital IC are also discussed. Electromagnetic interference (EMI) problems arising in connectors and vias are reviewed extensively, including novel studies of electromigration in VLSI.

In the analog domain, Volume 3 also addresses many subjects. This section starts with the basics of noise calculations for operational amplifiers. Included here is a review of fundamentals of circuit design using operational amplifiers, including internal noise sources for analysis. As an extension concerning noise issues in operational amplifiers, the material in this volume focuses on the very important subject of analog-to-digital converters (ADCs). In this area considerable effort is dedicated to proper power supply decoupling using bypass capacitance. Other noise issues in high-performance ADC are also addressed, including the proper design of switching power supplies for ADC, and the shielding of cable and connectors. Finally, at the IC level, work is included for studying RFI rectification in analog circuits and the effects of operational amplifiers driving several types of capacitive loads.

We end this volume with the study of system-level interference issues, such as intermodulation distortion in general transmitters and modulators, and the subject of cross modulation. This is followed by the concept of phase-locked loops (PLL) design, development, and operation. Because of the importance of

PLL in communications electronics, considerable space is devoted to the study of noise concerns within each of the components of PLL. Finally, Volume 3 ends with an attempt to explore interference at the level of transistors and other components.

Chapter 1 | Base Stations, Mobile RF Communication Systems, and Antenna Interferences

1.0 Introduction

Mutual interference in today's telecommunication systems is directly related to the International Telecommunication Union (ITU) frequency allocation laid down by that institution. There are several types of mutual interference among frequency-sharing systems: (1) interference between terrestrial stations; (2) interference between satellite–earth links; and (3) interference between terrestrial stations and earth stations.

As for interference between terrestrial stations (Noncellular, non-PCS, nonsatellite), in the lower part of the frequency spectrum (<200 MHz) most of the terrestrial services do not suffer from interference problems. Mutual interference that could exist has been limited to acceptable levels by good frequency planning. At higher frequencies (above 200 MHz) there is more spectrum for new services; space communication services such as PCS and cellular are important above 1 GHz. The demand is constantly increasing for more ground communication systems. At these frequencies, fixed services (i.e., services using ground terminals at fixed positions, such as cellular and PCS) normally use radio-relay or cellular networks in order to overcome the limited coverage area of personal communications systems. These ground networks employ relatively high transmitted power because of possible fading due to occasional multipath effects and are therefore potential interferers for other services in the same frequency band. Although these station-to-station links use highly directional antennas that focus the beam in the forward direction, other stations may still be affected even if they are located outside the main beams. Broadcasting services are normally not exposed to interference from other stations, as most terrestrial broadcasting bands up to 1 GHz are not shared with other permitted services. Broadcast distribution in the UHF and VHF bands is usually carried out by networks of auxiliary transmitters similar to the radio relay network of the fixed services.

1.1 Interference between Satellite–Earth Station Links

In low-earth-orbit satellites operating at frequencies above 1 GHz, there is a need to hand over communications from one satellite to another to maintain permanent

1

connections between earth stations and between earth stations and mobile personal communications services. In geosynchronous satellites, orbit utilization is limited because interference can restrict the allowable minimum spacing between neighboring frequency-sharing satellites, as shown in Figure 1.1. Additional satellite spacing is needed to take into account the small but inevitable perturbations of the satellites around their nominal orbital positions.

In order to prevent more serious interference problems, the ITU has prescribed several requirements for all satellite communication systems. The maximum power flux densities on the earth surface from transmitting ground stations and the sidelobes of transmitting and receiving earth stations have been limited. Operation of broadcasting satellite transmitters is urged to confine the radiation to the intended coverage area and reduce the radiation into other areas by means of multiple beam antennas or contoured antennas. The improvement of antenna systems for earth stations is one of the most obvious ways to meet these requirements. It is evident that the planning of a new satellite service in a frequency

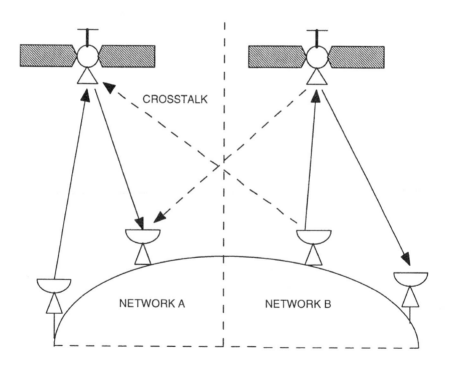

Figure 1.1 Interference from one satellite network to another.

band shared with another service will have to include the levels of mutual
interference between these two systems.

1.2   Interference between Broadcasting Terrestrial Stations and Satellite Earth Stations

In many satellite services above 1 GHz there is a need to share frequency bands
with terrestrial services. For example, the 4- and 6-GHz bands for the fixed
satellite service have to be shared with the fixed service. The same is true for
the 11- and 14-GHz fixed satellite service and the 12-GHz broadcasting satellite
service. The typical mutual interference situation (shown in Figure 1.2) is due

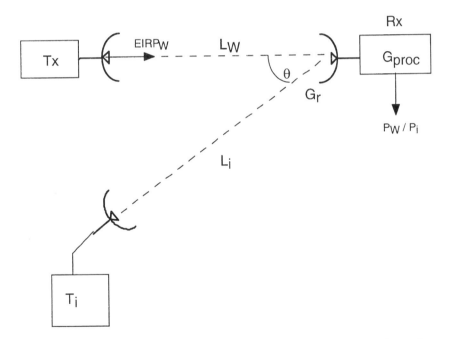

Figure 1.2   A general interference scenario, where P is the signal power at the receiver
input; $G_r(\theta)$ is the receiver antenna gain at the angle θ off boresight; L
is the path loss between transmitter and receiver; EIRP is the equivalent
isotropically radiated power; G_{proc} is the process gain defined by the receiver
improvement of the signal to interference ratio; and i and w are the interference signal and wanted signal, respectively.

to the relatively low received power in satellite transmission and the relatively high gain of possible interfering antennas. The dominant equation is given by

$$\frac{P_w}{P_i} = \frac{G_r(0)}{G_r(\theta)} \frac{L_i}{L_w} \frac{EIRP_w}{EIRP_i} G_{proc} \geq \left(\frac{P_w}{P_i}\right)_{min}$$

Technical means of protection of radio channels are based on enhancing the protection ratio P_w/P_i. The preceding equation shows that an increase in any of the following factors improves the interference immunity of the desired channel:

1. Antenna discrimination: $G_r(0)/G_r(\theta)$. The receiving antenna acts as a spatial filter and may therefore discriminate between wanted and interfering signals, provided that these signals arrive from different directions ($\theta = 0$). In addition, the antenna may discriminate between signals with different polarizations.

2. Propagation control = L_i/L_w. The path loss L_i of the unwanted signal can be increased by the introduction of obstacles on the propagation path in order to cause extra diffraction losses. This possibility, known as site shielding, is often inherently available in urban areas.

3. Superior transmitting stations: $EIRP_w/EIRP_i$. A brute-force method of suppressing interference effects in one own's system is simply to obtain extra equivalent radiated power, by increasing either the output power of the transmitter or the transmitting antenna gain.

4. Signal processing = G_{proc}. Often, signal processing will be the most effective and flexible technique for improving the system protection ratio. Processing can be carried out before, during, or after the demodulation process, depending on the system that has been adopted at the transmitter side.

Often in many communication channels we only have control of the receiver side of the radio channel. The nature of the transmitting sources is such that their properties cannot be easily changed. Furthermore, some of these interference sources come from unpredictable locations with variable powers and times on air. This makes systematic shielding by buildings unreliable and increases the difficulty of protection by antenna discrimination. One possibility for diminishing interference could be frequency diversity, such as switching to another transmitting frequency when one frequency is suffering from interference. However, this will mean a degradation of the wanted signal quality if the frequency choice for a desired program was optimum. On the other hand, site diversity has proven to be a useful principle to avoid interference. A second antenna can be situated elsewhere for the reception of the interference signal. Such an antenna should be positioned closer to the desired broadcast stations to improve the signal quality.

New technical means of protection, known as interference cancellation, have been developed and are able to discriminate and eliminate an interference signal from a wanted signal. The basic principle of interference cancellation is shown in Figure 1.3a. An auxiliary antenna pointed toward the interference source is used as a reference antenna to obtain a copy of the interference received by the main antenna. After complex weighting in amplitude and phase, the signal from the reference antenna is subtracted from the main antenna signal in such a way that the interference is eliminated. This static interference canceller is applicable whenever the direction of the interfering source is known and fixed, and different from the direction of the wanted source. Static systems are simple and inexpensive. However, an adaptive interference canceller as shown in Figure 1.3b is more suitable where the complex weighting is controlled by the output of a correlator, which compares the output signal of the system and the reference signal. The correlator adjusts the phasor modulator in such a way that the correlation between the input signals of the correlator is minimized, which is the case when there is no interference present in the output (provided that the reference antenna receives a clean interference signal). Adaptive interference cancellers can suppress unwanted signals for up to 50 dB. The block diagram of a more complex interference canceller is shown in Figure 1.4. The heart of the system is the complex phasor modulator (CPM) which controls the RF input signal, both in amplitude and in phase, depending on two DC control signals. The control signals are derived by correlating both quadrature components of the reference signal with the output signal, by means of two synchronous detectors.

The output signals of these detectors are integrated and then fed into the control inputs of the CPM (control phasor module). An automatic gain control (AGC) is used to guarantee a constant level of the reference signal at the inputs of the detectors. The IF bandpass filters are needed to define the channel to be protected; this can be varied by adjusting the local oscillator. The RF bandpass filters protect the broadband amplifiers against saturation by signals outside the frequency band to be protected. These filters must be changed or returned whenever the local oscillator frequency is changed. Figure 1.5 shows a typical result at baseband. The wanted signal is an unmodulated carrier, while the wanted signal is FM-modulated on the same RF frequency of 100 MHz with a power level at the output larger than the signal in the absence of the canceller.

In general, the process gain of the system will drop when the reference signal is not quite clean (that is, it contains some unwanted signal), because of correlation with the wanted signal. This implies the need to use very selective antennas.

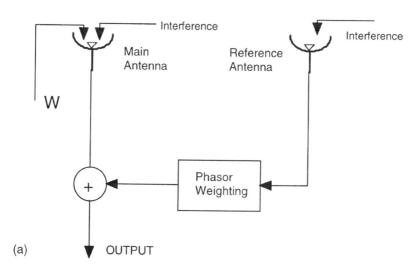

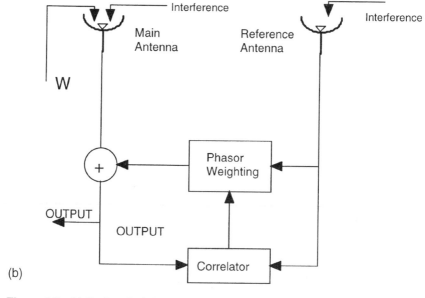

Figure 1.3   (a) Basic principle of interference cancellation. (b) Adaptive interference canceller.

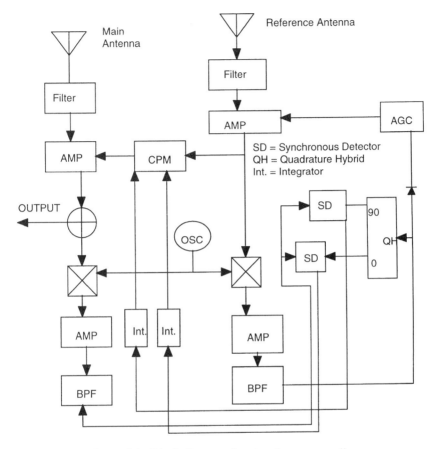

Figure 1.4   Block diagram of an interference canceller.

Because of the unpredictability and variety of the interference sources, the reference antenna should be sensitive to all horizontal directions, but the antenna should not receive a signal from the boresight direction of the main antenna. Because the canceller needs a clean reference signal to be correlated with the output signal, therefore, the ideal radiation pattern is then a circular symmetric pattern with a very sharp null in the direction of the wanted signal. To realize this minimum, the directivity of the main antenna can be exploited: subtract a suitable fraction of the main antenna signal from the transmitted power of broadcasting stations. The risk of interference from such broadcasting terrestrial stations into receiving earth stations is normally higher than that of harmful interference from transmitting earth stations into terrestrial stations. Therefore, although the

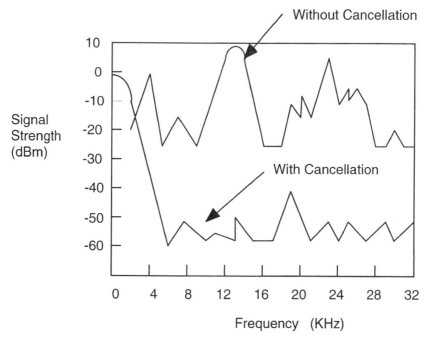

Figure 1.5   Baseband spectra of unprotected and protected unmodulated carrier due to interference from co-channel FM interference.

latter interference problem cannot be ignored, the former interference scenario causes more problems because typical multidestination satellite systems involve more receiving earth stations than transmitting earth stations. (See Figure 1.6.)

In addition to the general technical requirements imposed to limit the number of interference occurrences previously mentioned, the ITU has restricted the permissible horizontal radiated power of both broadcasting earth stations and radio delay stations, operating in shared frequency bands. In addition, a minimum elevation angle of 3° is prescribed for transmitting broadcasting earth stations. These limitations are not sufficient to prevent interference, since this interference phenomenon is unwanted propagation along a terrestrial path. This type of inter-ference can very well be enhanced by atmosphere or terrestrial influences.

1.3   Technical Protection against RF Interference

In the interference scenario shown in Figure 1.2, in general adequate protection of the wanted radio channel against RF interference from an unwanted source

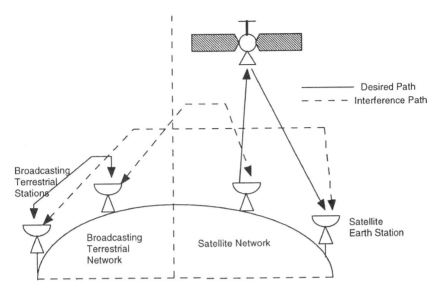

Figure 1.6   Interference between satellite networks and broadcasting terrestrial network.

requires a minimum value of the signal-to-interference ratio of the output of the receiver,

$$\frac{P_W}{P_i} = \frac{G_r(0)}{G_r(0)} \frac{L_i}{L_W} \frac{\text{EIRP}_W}{\text{EIRP}_i} G_{\text{proc}} \geq \left(\frac{P_W}{P_i}\right)_{\text{min}}. \tag{1.1}$$

Figure 1.7 shows the output signal of an omnidirectional antenna in such a way that the wanted signal is eliminated in the combined output.

If we use a system interference canceller of the type shown in Figure 1.4 and refer to the more simple Figure 1.7a, we can derive the formulas for the signal to interference plus noise ratios that can be obtained with or without the use of the cancellation system. The formulas are given by

$$(\text{S/IN}) = (\text{S/N})_0 \frac{1}{1 + (\text{S/N})_w G_1^2(\theta)}$$

$$(\text{S/IN}) = (\text{S/N})_0 \frac{1 + (\text{S/N})_w G_2^2(\theta)}{1 + (\text{S/N})_w G_2^2(\theta) + (\text{S/N})_w G_1^2(\theta)}$$

where

$G_1(\theta)$ = voltage gain of the main antenna in the direction of θ
$G_2(\theta)$ = voltage gain of the reference antenna in the direction of θ

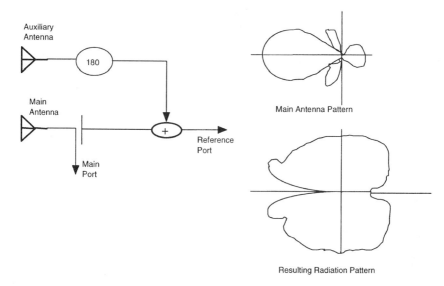

Figure 1.7   Obtaining the main antenna pattern of a wanted signal.

$(S/N)_i$ = signal-to-noise ratio of interference signal

$(S/N)_w$ = signal-to-noise ratio of wanted signal

$(S/N)_0$ = $(S/N)_i \, G_1^2(0)$ is the signal to noise ratio at the output in the absence of interference

(S/IN) = signal to interference plus noise ratio

1.4   Pager Antennas and Their Performance

One of the most important parameters in pager design is that of antenna performance. In Figure 1.8 we see a typical pager receiver using a double conversion approach.

The sensitivity of the receiver is limited by the system's gain and noise figures. In a noise-figure-limited condition, the carrier-to-noise ratio (C/N) is not high enough to provide an acceptable bit-error-rate (BER) on the received data stream. In order to achieve successful reception, the following two requirements must be satisfied:

$$C_r G \geq C_{if \, (min)}$$

$$\frac{C_{if}}{N_{if}} = \frac{C_R}{N_R} \times \frac{1}{F} \geq \left(\frac{C}{N}\right)_{IF \, min}. \tag{1.2}$$

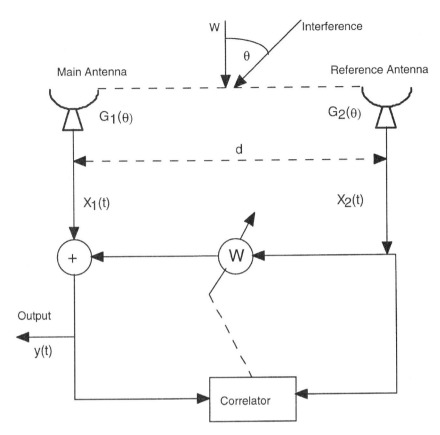

Figure 1.7a   Adaptive interference canceller using an ideal reference antenna.

where,

C_r	=	received signal carrier level
G	=	pager receiver front-end gain
$C_{if,min}$	=	minimum IF carrier level for FM detection
N_R	=	received noise level
F	=	IF signal level
N_{if}	=	IF noise level
$(C/N)_{IF,MIN}$	=	minimum IF (S/N) ratio for acceptable BER detection of the IF IC

It can be observed from this equation that there are two ways to maximize the received signal. One is to increase to gain G, which is not an easy task given the limited voltage (about 1 V) and current (1.5 mA) under which a pager

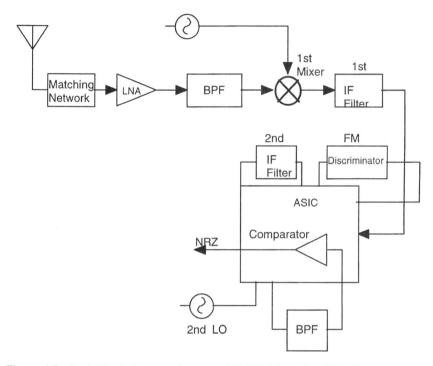

Figure 1.8   Basic block diagram of a pager. (Modified from Ref. [2] with permission
from *Microwaves & RF.*)

operates. The term N_R deals mainly with KTB noise, since pagers are often
placed in a tapered TEM cell for sensitivity evaluation and other forms of testing.
The most effective way to maximize the received signal is to maximize C_r and
minimize F. One way to optimize C_r is by optimizing the receiving loop antenna
structure. Most pager antennas are loop antennas. Loop antennas are generally
constructed with either round wires or flat strip conductors as shown in Figure 1.9. The flat strip design usually offers superior performance.

The signal power that an antenna can supply to the LNA is given by

$$P = \left(\frac{\lambda^2}{4\pi}\right) D \left(\frac{|E|^2}{Z_c}\right),$$ (1.3)

where

E = free-space electric field strength
Z_c = characteristic impedance of free space
D = antenna directivity

$$D = D_o \varepsilon K,$$ (1.4)

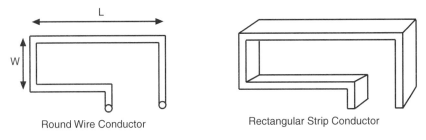

Round Wire Conductor   Rectangular Strip Conductor

Figure 1.9   Antenna configurations used in pagers.

where

D_o = directivity of a lossless loop antenna = 1.5 for small loop (similar to an infinitesimal electrical dipole)

K = mismatch loss between the antenna and the receiver input

ε = loop antenna efficiency

This means that for a given E and Z_c, the most practical way to increase the antenna sensitivity is to increase the loop antenna efficiency. For a single loop,

$$\varepsilon = \frac{R_r}{R_r + R_{loss}}, \tag{1.5}$$

where

R_r = loop radiation resistance = $20 \log \left[\left(\frac{2\pi}{\pi} \right)^2 A \right]^2$

A = loop area

R_{loss} = $R_s \, L/P$

R_s = high frequency resistance of conductor = $\sqrt{(\pi f \, \mu/\sigma)}$

L = length of loop

P = cross-sectional perimeter of the conductor

σ = copper conductivity (5.813×10^7 s/m)

μ = magnetic permeability of loop conductor ($4\pi \times 10^{-7}$ A/M) for copper

Flat strip conductors are preferred over round conductors because of their lower skin effect loss due to the higher ratio of surface area to cross-sectional area. It turns out, if calculations are performed, that the efficiency of a loop antenna using a rectangular strip is about 2 dB better than that of round-loop antennas, even if both conductors have the same cross-sectional area. The method of moments [1] can be used to calculate the radiation field pattern in free space for both a round wide loop and a flat rectangular strip (Figure 1.10).

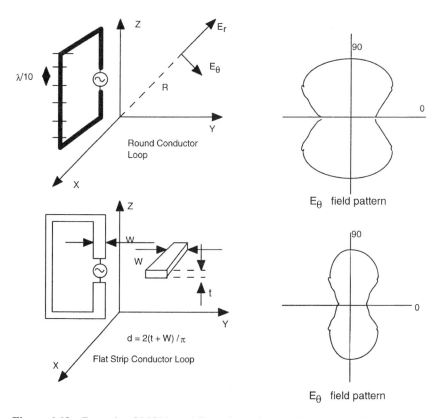

Figure 1.10   Example of MOM modeling of round and strip conductor loop antennas.

In using the method of moments for the flat rectangular strip, we have transformed the rectangular cross-section of the rectangular strip into its equivalent round structure. The efficiency of a multiturn loop antenna is given by

$$\varepsilon_m = (\varepsilon_1 + 10 \log m) \quad \text{(dB)}; \quad (1.6)$$

where m is the number of turns, and ε_1 is the efficiency of a single loop antenna. The associated radiation resistance and associated loss resistance for multiloop antennas are given by

$$R_m = m^2 R_r \quad (1.7)$$
$$R_{\text{loss},m} = m R_{\text{loss}}.$$

It seems that increasing the number of loops would increase the antenna efficiency. However, because these loops are placed in very close proximity, the current is not uniformly distributed around the wires, but depends on the external skin effect and the proximity effect. The proximity effect causes the current to be in the outside edge of each conductor in the same manner as the skin effect forces the current to the outside surface of the conductor. Therefore, for m parallel wires, the total resistance per unit length is given by

$$R_T\left(\frac{\Omega}{m}\right) = m\left(\frac{R_s}{2\pi a}\right)\left(1 + \frac{R_p}{R_o}\right) \tag{1.8}$$

and the loss resistance is given by

$$R_{loss} = MR_{loss}\left(1 + \frac{R_p}{R_s}\right), \tag{1.9}$$

where

R_s = surface resistance
R_p = resistance due to proximity effect
R_o = $mR_s/2\pi a$ = the skin effect resistance per unit length (ohms/m)
a = conductor radius
R_{loss} = the loss resistance for an m-loop antenna with the proximity effect included

It has been shown [2] that the antenna efficiency for a multiloop antenna can be given by

$$\varepsilon_m = \varepsilon_1 + 10 \log m - 10 \log\left(1 + \frac{R_p}{R_o}\right). \tag{1.10}$$

As can be observed from this formula, the radiation resistance is increased by using a several-double-loops configuration, but only up to a limit, because the overall efficiency can actually decrease if loss resistance terms increase, which can be caused by longer wire length, the proximity effect (R_p), and the reduction in surface area caused by the narrower strip width.

If pager frequencies were to increase, then because of the pollution of the UHF band at 150 MHz, another problem that must be dealt with is the self-resonance frequency of the antenna, which can pose difficulties in obtaining practical matching networks. The equivalent circuit of a loop antenna is shown in Figure 1.11 where the loop inductance L_p is given by

$$L_p = (m^2\mu_o/\pi)\left\{L\ln\left(\frac{2L\omega}{a(L + L_c)}\right) + \omega\ln\left(\frac{2L\omega}{a(W + L_c)}\right) + 2(a + L_c - (L + W))\right\},$$

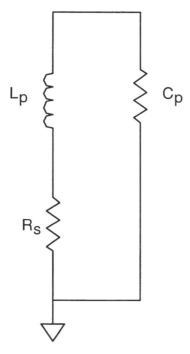

Figure 1.11   Circuit representation of a loop antenna.

where

$L_c = (\omega^2 + L^2)^{0.5}$, L and W are defined in Figure 1.9,

and C_p is the stray capacitance,

$C_p = 1/(2\pi f_r)^2 L_p$,

where f_r is the antenna's resonance frequency (the frequency at which the output impedance becomes resistive). The term f_r is usually found experimentally. However, f_r can also be found analytically using the method of moments, though the results could be somewhat different from experimental data as a result of stray capacitances in the equipment setup for multiple loops. A matching network (L_m, C_m) such as that shown in Figure 1.12 can be used between the loop antenna and the LNA input (R_{in}, C_{in}). The terms C_{m1} and C_{m2} are capacitance for the matching network.

Two-way paging, which means combined two-way voice/data, is increasing in popularity. Such a pager is shown in Figure 1.13. Notice the presence of a

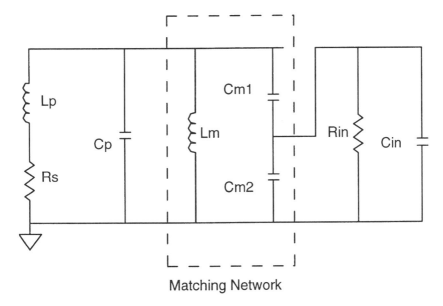

Figure 1.12 Matching network for a loop antenna.

software interface to host a microprocessor (often with a DSP IC) which would make it easier for designers to embed the chipset and setup of a ReFLEX protocol system without special in-depth knowledge of the ReFLEX protocol itself.

1.5 Cellular Phone Structures

The design of cellular phones requires the proper selection of integrated circuits among a wide selection of possible candidates. The design process is aided by three existing but different true time-division multiple access (TDMA) standards: EIA/TIA IS-54, which regulates cellular service in North America; ETSI-GSM, which regulates cellular service in Europe and parts of Asia; and RCR std 27B, which regulates cellular services in Japan. The North America standard accommodates existing analog cellular systems as well as the digital formats; the GSM and PDC standards define only purely digital systems.

A typical cellular phone consists of both baseband and RF as shown in Figure 1.14. The baseband section performs all of the audio processing command and control and analog/digital conversion. The RF section contains all the transmitter and receiver and supporting circuits.

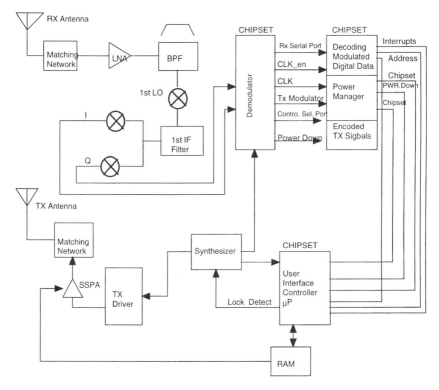

Figure 1.13   Block diagram of a possible two-way pager.

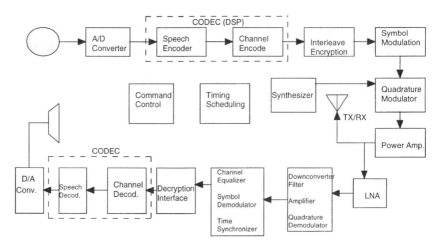

Figure 1.14   Basic block diagram of a cellular phone.

The audio-to-cellular interface consists of a speech coder–decoder (codec) which receives the audio signal via a dedicated low-noise amplifier and provides audio signal to a transducer through another power amplifier. Independent input and output amplifiers are required to support full two-way operation, which allows both parties to transmit to each other simultaneously. The amplifier must meet the SNR and harmonic distortion requirements of one of the given standards. The output voltage gain of the amplifier must be matched to the input requirements of the analog-to-digital converter (ADC).

In an analog phone, the output of the microphone amplifier is routed through a switch to an audio filter. The analog op-amp filter bandwidth is limited to 3 kHz, but it preemphasizes higher frequencies to improve modulation efficiencies. In digital phones the microphone amplifier's output is routed to an audio A/D converter. The audio converter digitizer samples it at an 8-kHz rate. The output of this ADC is buffered by a digital signal processor (DSP) for speech coding and further pretransmission processing.

An incoming signal from the receiver is uncoded in the DSP and routed to an audio digital-to-analog converter (DAC). The analog audio is sent to a power amplifier that drives the output of a transducer. Signal processing is handled by a DSP operating with algorithms tailored to the requirements of the cellular standard for which the phone is used. The DSP performs speech coding and modem functions. Depending on the phone's architecture, the DSP may also perform channel coding and encryption. The DSP used in a given phone must be designed specifically for the standard that the phone will use. GSM, PDC, and IS-54 employ TDMA for digital transmission. In TDMA, the channel capacity is divided into time slots. Each phone is sequentially and repetitively allocated a series of slots, which allows multiple phones to share the same channel simultaneously (Figures 1.15 and 1.16).

Channel coding minimizes the potential of digital loss bits by the process of convolutional encoding. The encoding adds redundant bits to the data stream, providing backup for the most significant bits so that if these are lost in transmission, the signal can be reproduced. Channel coding can be done in software using a DSP or in hardware using an ASIC.

Multipath RF signals where the primary signal and several images of the same signal (although at different power levels) arrive at the antenna is caused by reflections from obstructions with sufficient reflectivity at the RF frequency of interest. Multipath signals interfere with each other constructively or destructively. In destructive interference, the signal loses radiated power or the bit-error rate (BER) increases.

In order to decrease burst errors, interleaving is used to spread short or long lengths of decoded information bits with a BER approaching 0.5 into the data

Frame = 40 msec

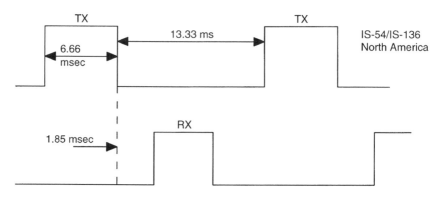

TX-RX frequency range: 45 MHz
Max. inclusive channel change: 70 MHz

Figure 1.15   TDMA architecture in cellular phones.

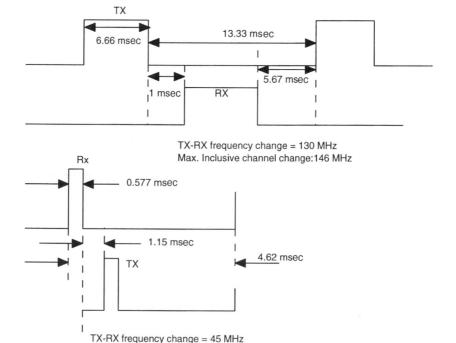

Figure 1.16   Further analysis of the TDMA architecture.

stream where the decoded information BER is much lower. To minimize cell telephone identification and verification data being decoded by an unauthorized party, the data on the RF channels can be encrypted. The GSM standard contains options for the encryption of messages and traffic channels. IS-54 does not, but IS-136 does provide the capability for encryption. Encryption is performed by applying exclusive-or logic between the digital data stream and a separately generated cipher stream, allowing ciphering or no-ciphering options. The cipher-stream generating algorithm can be performed in software by the DSP or using ASIC. However, this cipher algorithm is usually more complex than the channel coding and interleaving algorithms, and as a result ASIC hardware design is typically used to perform the cipher stream generation while the DSP is used for exclusive-or combination of the digital and cipher streams.

The microcontroller in a cellular phone does the command and control functions. Included are the interpretation of functions received by the base station and the generation of commands to the relevant circuits. It also continuously adjusts the telephone's transmitted power to maintain the required signal strength by the base station. An important function is ensuring that the phone transmits and receives data only during the assigned intervals in the TDMA cycles. In the IS-34 standard cellular phone, the microcontroller enables appropriate analog or digital-mode specific portions of the transmit and receive signal.

The baseband signal composed of encoded, interleaved, and unencrypted digital data streams is routed to the RF transmitting section, which performs symbol modulation, quadrature modulation, and power amplification. The symbol modulation converts the digital data stream into the differential phase-modulation formats required by each standard: GSM requires Gaussian minimum shift keying (GMSK), whereas IS-54/IS-136 and PDC require TI/4 differential quadrature phase-shift keying (TI/4-DQPSK).

In the RF section the symbol modulator outputs two orthogonal analog signals: the in-phase (I) signal and the quadrature (Q) signal. The symbol modulation provides necessary spectral filtering. The main act of the phone's transmitter is its quadrature modulation, which converts the I and Q analog signals from the channel code into an RF signal for transmission. Quadrature modulators can be implemented in a single IC designed to a specific standard. In some cases a direct modulation technique is preferred in order to reduce filtering in the transmit path. Direct modulation also eliminates the need for an upconversion mixer, reducing external components and power consumption. Some external filtering is always required to avoid spurious transmission and noise in the receiver.

The transmitting channels are spaced from the receiving channels by the frequency specified in the standard (45 MHz in IS-54 and GSM) and 130 MHz for the Japanese standard. A single synthesizer is used most often supplying a frequency. The transmitter carrier frequency can be generated by mixing the UHF synthesizer's output with the output of a fixed frequency oscillation (see Figure 1.17).

An audio signal can be transformed into an I and Q format in baseband processing and used to produce FM in the quadrature modulator, simplifying the design of the VCO.

RF signals generated by the modulator are routed to the power amplifier. Two or more preamplifier stages between the modulator and power amplifier may be used. The output of the power amplifier is routed to the antenna with the use of a diplexer, which provides isolation between the transmitter output and the receiver input.

In a TDMA system an active telephone is assigned a specific periodic time slot during which it can transmit on its assigned channel. Furthermore, the phone receives information only during its assigned time slot on its receiving channel. In order to minimize crosstalk within the cellular phone, the transmit and receiver slots occur at different times so that the phone is either transmitting, receiving, or in standby mode.

Signals transmitted by the base station are intercepted by the telephone antenna and routed through the duplex via the receiver's low-noise amplifier. The output of the amplifier is routed through an image filter, a surface acoustic wave (SAW) device, that passes signals within the cellular receiver band while attenuating leakage and noise generated at frequencies outside the cellular band. The signal is routed from the image filter to the RF mixer and is multiplied by the output of a local oscillator to produce the first IF. A single variable-frequency synthesizer

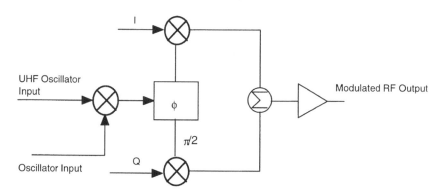

Figure 1.17   Modulated output generation from UHF input.

can reduce the compared count. In GSM and IS-54, the transmit frequency can be generated by heterodyning the variable-frequency UHF oscillator with a fixed offset oscillator (the sum of 45 MHz and the desired RF), as shown in Figure 1.18.

A discrete circuit IC can be used to implement the mixer; however, this can bring the cost up significantly.

Most cellular phones use the double conversion receiver architecture (Figure 1.19). In this approach the RF mixer's output contains the first IF, typically in the range from 45 to 130 MHz. The first IF is selected from the broadband output of the RF mixer by using a narrow IF filter (quartz SAW filter). The first IF signal is routed into a second mixer, to produce the second IF at a much lower frequency (450 kHz to 10.7 MHz). This second mixer output signal is routed through one or two second IF filters, usually ceramic, and amplified. The amplifiers can be a discrete circuit on a pad of a single IC.

The final output signal is converted to audio by a discriminator and is routed through filtering and processing to an audio power amplifier that drives the output transducer.

For digital signals, the second IF amplifier is made up of a strip of IF amplifiers equipped with automatic gain control. The gain of these amplifiers is adjusted to maintain amplifier linearity. After amplification, the IF signal is routed to a quadrature demodulator. The detected I and Q signals are routed to the receive side of the channel coded A/D converters, which translate the symbols represented

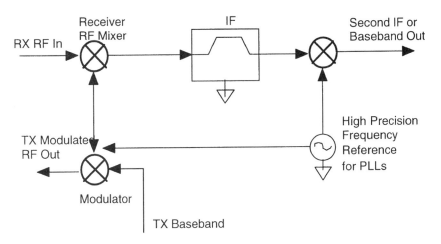

Figure 1.18 Generation of the transmitting frequency.

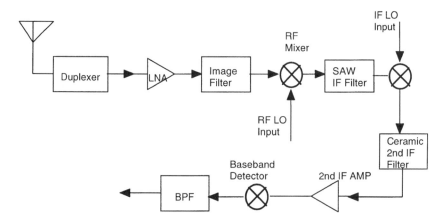

Figure 1.19   Double conversion receiver architecture.

by the I and Q signals into digital signals that are routed to the receive side of the DSP. The DSP decompresses the received data into a digital representation of the speech transmitted by the caller. The digital signal is sent to the receiver side of the speech codec, where its DAC synthesizes an analog signal representing the caller's speech. This signal is amplified and sent to the transducer for hearing.

A more detailed diagram of digital cellular phones can now be constructed (Figure 1.20).

1.6   Base-Station Antenna Performance

The antenna performance of ground stations can be characterized in terms of radiation pattern, gain beamwidth, and impedance bandwidth. These parameters are applicable to both transmit and receive functions, since this is valid from the reciprocity symbol. In the process of defining antenna performance in terms of radiator pattern, the important parameters to consider are gain at different coverage angles, sidelobe and backlobe levels, and polarization response. A polar pattern approach shows an accurate illustration of RF energy in three-dimensional space.

The radiation pattern coverage of a typical antenna in polar coordinates is shown in Figure 1.21.

For example, consider the plane cuts of a vertically polarized dipole antenna. The θ cut is directional with characteristic nulls at the top and bottom. The ϕ cut (azimuth) is omnidirectional in the horizon. The polarizations are often described as E_θ for vertical, E_ϕ for horizontal (Figure 1.22).

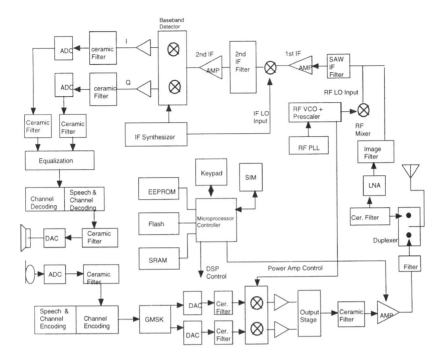

Figure 1.20 Detailed block diagram of a digital cellular phone.

Antennas are directional by nature and they are more likely to radiate in certain directions. Antennas are designed with controlled directive properties to guide the available energy into certain desired coverage sectors. The directivity multiplied by antenna efficiency is the gain (dB), expressed relative to a reference source such as a hypothetical isotropic source that radiates uniformly over a spherical surface. The gain of an isotropic source is unity, or 0 dB. (See Figure 1.23.)

Other parameters of interest include the antenna beamwidth and beam area. The beamwidth is the angular width to either side of the peak of the beam where the single lobe is at the half-power point (-3 dB). Beam area is the product of the principal plane beam width in degrees squared (i.e., $\phi \times \theta$).

For a lossless antenna, the gain is defined as the directivity,

$$D = \frac{41,253}{\theta \times \phi} \tag{1.11}$$

where D = directivity (unitless or $10 \log D \equiv$ dBi).

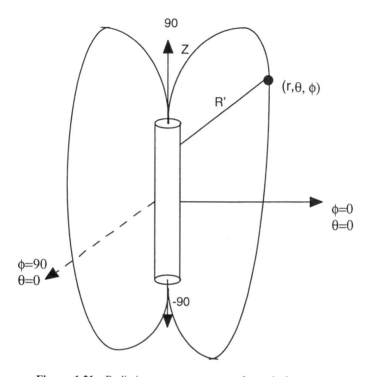

Figure 1.21   Radiation pattern coverage of a typical antenna.

Directivity multiplied by antenna efficiency yields absolute gain, and a number more often used is given by the expression

$$D = \frac{30,000}{\theta^0 \times \phi^0}. \tag{1.12}$$

All antennas have an effective aperture (A_e), and the gain can be related to the effective aperture and given by the expression

$$G = \frac{4\pi A_e}{\lambda^2}. \tag{1.13}$$

The operating impedance bandwidth for an antenna is defined by a maximum acceptable value of voltage-standing-wave ratio (VSWR). The maximum VSWR of base station antennas is 2:1 over the operating bandwidth. It is important to have properly sized and matched base-station antennas because of the risk of damage from reflected high transmitted power.

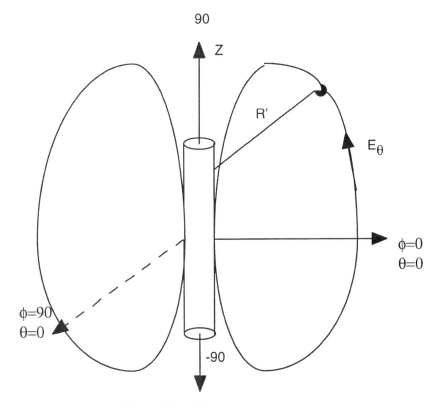

Figure 1.22 Polarization representation.

1.6.1 SUITABLE ANTENNAS FOR BASE STATIONS

Antennas that are suitable for use in wireless systems base stations fall into three basic groups: resonant, aperture, and phase array. Most resonant antennas used in wireless communications are either $\frac{\lambda}{2}$ monopole patches or arrays of these elements. Elements of cylindrical shape are mostly used because of symmetry-impaired bandwidth and mechanical stability as shown in Figure 1.24.

Variants of the basic dipole and monopole design include the folded dipole, cylindrical and biconical dipole, and conical monopoles. Biconical geometry is of interest when large bandwidths are required. One of the main advantages of the dipole is its 360° horizon coverage.

Aperture antennas are directional antennas when the radiated energy is focused in a given direction. They use a single element to illuminate a reflector and

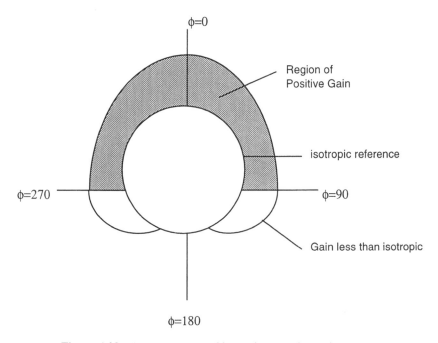

Figure 1.23   Antenna pattern of isotropic vs nonisotropic antenna.

associated directional components. Most often, this illuminator is a dipole. Among the reflectors most often used are the plane and corner reflector. Yagi-Uda antennas are also used as reflectors, but their gain drops off quickly with an increase in the number of reflector elements. Corner and plane reflectors offer advantages over other antennas in terms of pattern and gain control.

These types of antennas are expanding rapidly to meet electrical and mechanical demands of wireless systems. Samples of arrays that are useful in base stations include collinear arrays of dipoles, dipole arrays over a plane, and microstrip patch arrays (Figure 1.25).

The planning of a base-station antenna is a task that involves the shape of the coverage area, the location of the site, and the antenna type with the most suitable azimuth, as shown in Figure 1.26. Antennas such as single collinear dipoles and elementary reflectors of four separate reflector antennas provide azimuth patterns suitable for square coverage areas.

Antennas offering an azimuth pattern suitable for rectangular sites include multiple element reflectors and phase dipole pairs. Irregularly shaped coverage areas can be treated as rectangles or squares. The dimension of the covered area

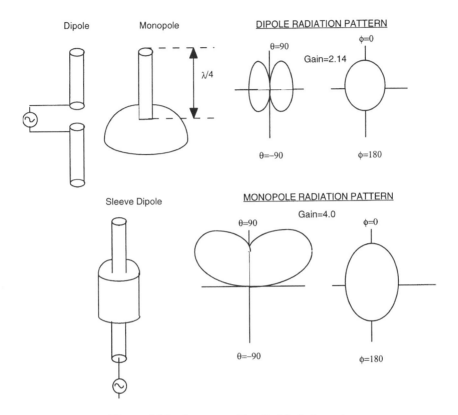

Figure 1.24 Antennas with cylindrical elements.

and height of the site are needed to determine the azimuth and elevation beam-width requirements if we apply the beam area formula using the expression

$$G(\text{dBi}) = 10 \log \frac{29,000}{\theta^0 \times \phi^0}, \qquad (1.14)$$

where θ is the elevation plane pattern and ϕ is the azimuth plane pattern in degrees. The gain and beamwidth are important requirements early in the design. Once the basic requirements have been obtained, the more specific antenna gain can be obtained from the expression

$$G_t = \frac{P_r(4\pi R)^2}{P_t G_R \lambda^2}, \qquad (1.15)$$

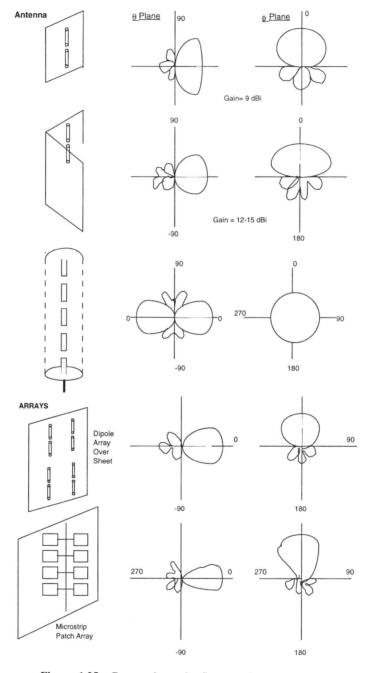

Figure 1.25 Commonly used reflector and array antennas.

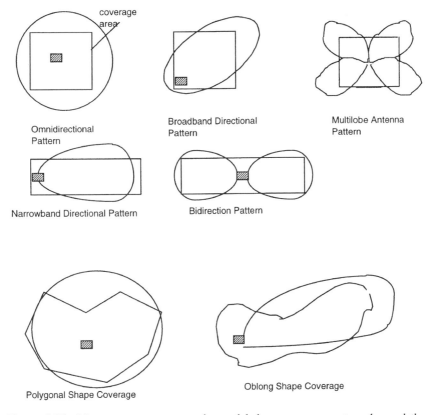

Figure 1.26 Most coverage areas can be modeled as square or rectangular, and the appropriate antenna type must be used. (From Ref. [3] with permission from *Microwaves & RF.*)

where

G_t = site antenna gain
P_R = remote receiver minimum discernible signal level
P_t = transmitter power
G_R = remote receiver antenna gain
R = range in feet or meters

In terms of decibels, the gain can be expressed as

$$G_t(\text{dBi}) = 10 \log P_R + 20 \log (4\pi R) - 10 \log P_t - 10 \log G_R \\ - 20 \log \lambda - 10 \log K_L, \tag{1.16}$$

where K_L is the loss, including free-space path loss.

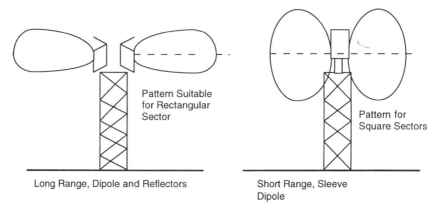

Figure 1.27   Antenna types/patterns suitable for rectangular and square sectors.

The term K_L must account for all diffraction and absorption losses. If operating in an adjacent or near channel of a given site, the possibilities of interference are a real possibility. Interference can be due to receiver desensitization in strong RF fields, spurious transmitter emissions, or receiver response to spurious inputs. Isolation can be achieved by using antenna decoupling. Even though the directive and polarization properties provide a certain level of decoupling, the amount of coupling can be obtained to a first approximation by using the expression

$$\text{Coupling} = \frac{P_2}{P_1} = \left(\frac{\lambda}{4\pi d}\right)^2 G_t(\theta_1)G_r(\theta_2), \tag{1.17}$$

where the parameters of interest are given in Figure 1.28. and in Eq. 1.17.

$\lambda =$ wavelength in meters
$d =$ antenna-to-antenna spacing in meters
$G_t(\theta_1) =$ gain of the transmit antenna of angle θ_1 relative to isotropic level
$G_r(\theta_2) =$ gain of the receive antenna of angle θ, relative to an isotropic level
      of 1.0

In this expression we are assuming that no sidelobe interference is significant and no polarized coupling exists. However, all antennas exhibit areas of depolarization; it is possible to physically align co-polarized and cross-polarized sectors in order to obtain good decoupling results.

When we have azimuth patterns and elevation patterns, and gain data have been selected, the following information will improve isolation.

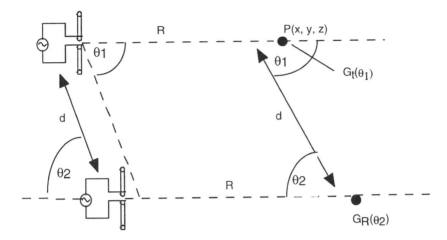

Figure 1.28   Parameters of antenna-to-antenna interference.

- If we are using omnidirectional antennas for square sectors and if such antennas have to be collocated, place such antennas on opposite sides of a tower structure. Adjust the antennas' spacing to obtain optimal cancellation without severely altering each antenna's coverage area.

- Install co-located omnidirectional antennas off-center and as far apart as possible.

- Place omnidirectional antennas with non-tilted patterns as high as possible in the tower relative to all other antenna types at the site.

- Place high-gain directional antennas back-to-back for maximum decoupling.

- Specify desired backlobe and sidelobe levels; for example, typical backlobes are in the -25 dB range relative to the peak level.

- Obtain out-of-band performance specifications of antennas for VSWR response.

We must be well aware of conducting structures. If an antenna is spaced more than 0.75λ from a large conducting surface, deep nulls are generated [3]. For antennas with small masts, acceptable pattern performance can be obtained by reducing the spacing to 0.70λ. For omnidirectional antennas near the shadow of large conducting structures, the solution is to raise such antennas above such structures.

Finally, it must always be remembered that conducting structures near the resonant length of a given antenna behave like the parasitic elements of an array, causing random nulls and lobes.

Structures that are nonconducting can also distort the radiation pattern of antennas. If the nonconducting structure is small in diameter or surface area, a spacing of 0.5λ or more can help eliminate antenna pattern distortions. Elevating omnidirectional antennas above larger nonconducting structures is also a preferred solution. With the possible exception of single-element dipoles, narrow elevation beamwidths of base station antennas are less susceptible to site interference. However, antennas placed at the center of a large roof area are prone to diffraction loss unless they are raised high enough for the main beam of the pattern to clear the roof edges [3]. If the height is a problem, such antennas should then be placed near the edge of a roof. Unidirectional antennas such as patch or dipole arrays can be located on either conducting or nonconducting surfaces without much of a problem. (See Figure 1.29.)

1.7   Environmental Factors in Antenna Selection

Wireless communications can be either bounded or unbounded where antennas are radiating either in an indoor environment (bounded) or radiating in free space. The RF energy launched from an antenna and which travels different terrains can be absorbed at different unintended locations. Furthermore, the signal can be reflected along the transmission path from the source antenna to a receiving antenna; this causes the signals to be received at different times, causing either constructive or destructive interference.

It must be remembered that the attenuating current at a given frequency which flows through an antenna element produces electric fields in the same direction as the radiating element and a magnetic field perpendicular to the electric field in far fields (Figure 1.30). The magnetic-field components concentrate about the antenna axis. The radiating antenna is always tuned to an operating frequency; the energy stored in the antenna, behaving as a resonator, is given to free space in the form of radiated energy.

The electric field orientation is the polarization sense of the propagated wave. The polarization is an important parameter in wireless design, since all antennas exhibit a principal E-field polarization alignment. An electric field is considered to have vertical polarization if it is perpendicular to the earth's surface and horizontal polarization if it is parallel to the earth surface. A circularly polarized wave is the resultant sum of two equal-amplitude electric field vectors in phase

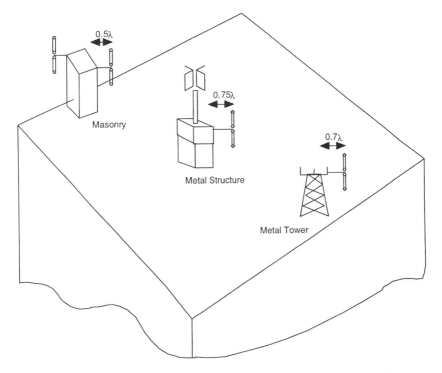

Figure 1.29 Proper location of different types of antennas depending on type of supporting structure.

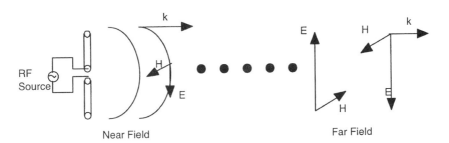

Figure 1.30 Illustration of near-far-field patterns of an antenna.

quadrature (Figure 1.31). A circularly polarized wave, therefore, contains all senses and all information of a linearly polarized wave with a reduced amplitude.

Polarization is very important in system performance, which is related to the polarization alignment between base-station and receiver sites. The maximum power occurs when the polarizations of transmitting and receiving antennas are similar.

Line-of-sight propagation spreading is one of the major electromagnetic propagation losses. As the RF signal radiates, it spreads and expands into a spherical surface. The available RF power is distributed over this surface and grows weaker with increasing range. From the inverse square law ($1/R^2$) the signal is reduced by 6 dB for every doubling of the distance from the source in the far field. The other path loss between point source radiators with spherical patterns can be computed by

$$L_{\text{path loss}} \text{ (dB)} = 36.6 + 20 \log f \text{ (MHz)} + 20 \log d \text{ (miles)}. \quad (1.18)$$

For ground losses, the solid material can be treated either as dielectric or as an imperfect conductor. Some factors that affect propagation of wireless systems include trees, mountains, buildings, and losses from the earth itself (Figure 1.32).

As a result of reflections, this signal travels many different routes from the transmitter to the receiver, causing an effect known as multipath distortion. The differences in signal path lengths create signals with different phase relationships that mix within the front end of the wireless receiver as shown in Figure 1.33. The velocity of RF energy is slowed with the passage of such a wave through dielectric materials that are more dense than dry air. This is refraction, which is the change in propagation direction shown in Figure 1.33.

Because of diffraction, energy is usually bounced around obstacles until it reaches its target, although at a reduced strength level. Most of this diffraction

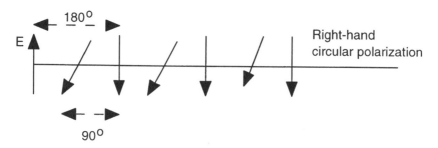

Figure 1.31 Illustration of antenna polarization.

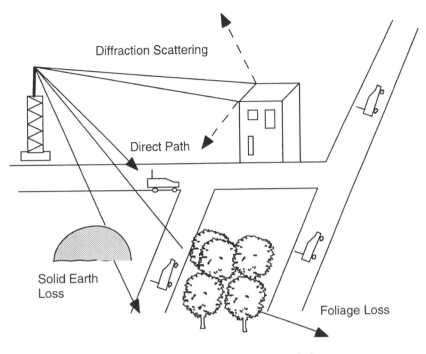

Figure 1.32 Factors affecting propagation in wireless systems.

is caused by edges which cause secondary radiation sources when illuminated. Because different linear polarizations reflect and penetrate surfaces differently, the most cost-effective way to reach inside structures is through the use of circularly polarized base-station antennas. Circularly polarized waves contain all senses of linear polarization and will always contain the most favorable electric field vector for any given incident angle. Scattering the energy in many directions gives the effect of some sort of energy loss, and this term is often called *diffraction loss*.

1.8 Performance of Dipole Arrays

The design of a dipole array must first start with the choice of the dipole in transverse dipole arrays. The feeding line and radiating elements are separated by a dielectric layer [4] as shown in Figure 1.34.

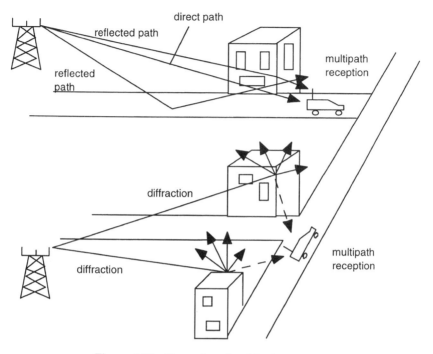

Figure 1.33   Illustration of multipath reception.

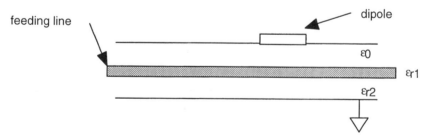

Figure 1.34   Feed element for a dipole array.

An advantage of this type of structure is that coupling phenomena can be controlled with a wide variety of parameters such as the dipole offset with respect to the feeding line Δy, the dipole length L_n, and dielectric layer heights h_1 and h_2. The design trade-off is that the input impedance of the elementary source will be dependent on these parameters. The array design will ensure the correct current distribution at every antenna element, as well as proper matching at the input port. This can be achieved using design equations based on work by Elliot and Stern [5]. Elliot and Stern proposed an array synthesis method for longitudinal dipoles that are electromagnetically coupled to a microstrip line. This technique incorporates a method that accounts for the coupling between elements. In this analysis the energy transfer is separated into two parts due to coupling between the feed line and the dipole and to mutual transfer between dipoles.

The analysis involves extracting each element from the array and relating the element to its feeder. The serial excitation of the array is neglected. Each dipole coupled to its line is represented by a four-part element referenced to a medium longitudinal plane. Any access of the bilateral four-port networks becomes one between the N ports of the multiport network which forms the array. Figure 1.35 illustrates each dipole as a bilateral four-port network. The dipole is described by the voltage V_n and current I_n. These parameters are related according to the equation

$$I_n = Y_n V_n, \tag{1.19}$$

where Y_n is the dipole's active impedance.

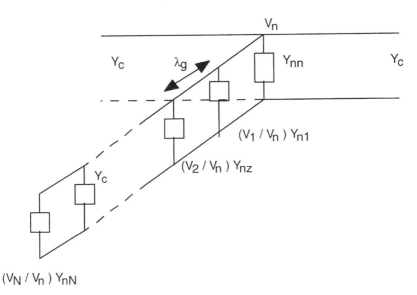

Figure 1.35 Electrical representation of a dipole array. (Modified from Ref. [4] with permission from *Microwaves & RF.*)

It can be shown that the method of moments can be used for calculating I_n and V_n from the expression

$$I_n / V_n = Y_{nn}C_n + \sum_{m=1,\ m \neq n}^{N} (I_{nm} / V_n), \qquad (1.20)$$

where the unknowns are given by Y_{nn}, C_n, Y_{nm}, and I_{nm}/V_n, and C_n is the ratio between the dipole and feeding line currents. The summation terms represent the mutual coupling between elements. Once the I_n distribution is obtained, the radiation pattern can then be calculated as shown in Figure 1.36. With optimization techniques, the array geometry $(L_n, \Delta y)$ can be obtained.

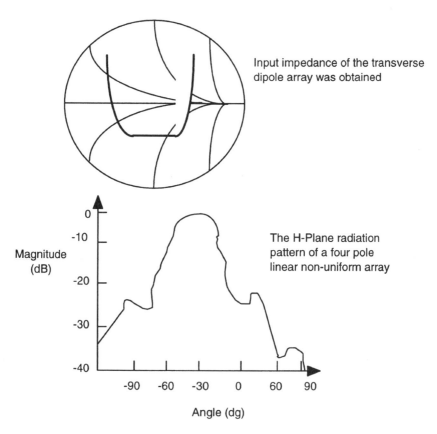

Figure 1.36   Radiation pattern of the dipole array geometry and input impedance.

1.9 Passive Repeater Technology for PCS Systems

Cell-based systems are having increasing difficulty in handling the great number of new subscribers. Advances in digital modulation and signal processing capabilities will increase capacity, but in reality this would be of no avail unless we develop well-confined coverage areas. These confined coverage areas will allow maximum use of limited spectrum assignments and will provide more reliable communications. One of the most important trends in ensuring good and reliable coverage is in the use of ''smart antennas''; however, smart antennas can be expensive and are ineffective in certain environments. It has been shown [6] that a passive repeater design could be placed in a covered zone in order to modify and shape the radiation so as to reduce the losses created by multipath, diffraction, and shadow regions. In reference [6a] this device has been named the space lattice passive repeater (SLPR). The SLPR can be described as a three-dimensional diffractive grating composed of multiple segments and then stacked metallic plates that are insulated from one another. The specifications for the spacing between each segment and the number of segments employed are designed to alter the linear polarization, redirect the path, and amplify the magnitude of radio-wave propagation. This enables the SLPR to extend RF propagation zones around objects such as buildings, doorways, corridors, and geological structures. The SLPR has application not only for long-distance signal enhancement, but also for indoor and urban wireless applications.

These passive repeaters fall into two major categories: reflector type or back-to-back antennas. The reflector can theoretically repeat a signal with gain, but without alteration in the wavefront polarization; the back-to-back antennas can alter the angle of polarization, but often with significant loss of signal strength. The SLPR combines positive elements of both approaches, and it can alter the path of propagation in more than one direction. The SLPR can redirect the original path of propagation into three major directions relative to the source antennas serving multipoint-to-multipoint applications. This capability allows the SLPR to be integrated into complex cell-based systems whose coverage areas meet at multiple junctions.

The construction of the SLPR is simple, as shown in Figure 1.37. This basic representation of the SLPR is a radical departure from what would be expected in a traditional passive repeater design. Instead of using conventional antenna and reflector technology, the design blends more into natural propagation phenomena: The parameters are an artificial approximation of an elevated tropospheric duct. These ducts have been studied for many years as effective natural conduits for RF energy. From Figure 1.37 it is apparent that as a wavefront approaches

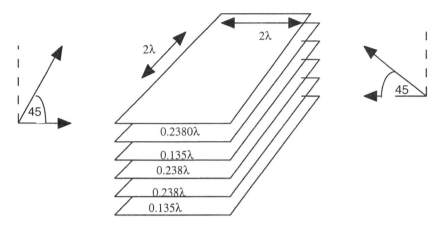

Figure 1.37 Physical representation of space lattice passive repeater. (From Ref. [6a] with permission from *Wireless Systems Design.*)

one of the four apertures, it passes through the structure with a 90° alteration in the direction of propagation. This effect is seen through both sides of the device, resulting in a "T" shape in the unit's radiation pattern. In addition, a component of the original wavefront is passed unaltered via the aperture. It is easily observed that the passive repeater has the ability to accept an incoming wavefront, redirect it in both orthogonal directions with a 45° alteration in polarization, and pass the original signal unchanged directly through the rear aperture without alteration.

The example in Figure 1.38 helps to show the effectiveness of the SLPR. The incoming RF arrives at an area resembling several buildings where multipath, shadowing, and diffraction are ever-present—for example, in the cellular signals throughout a building using a fiber-optic RF distribution system.

In the past, passive coaxial distribution systems required that a base station or repeater be connected to either a leaky coaxial cable or a tree-and-branch network with taps at each antenna location. However, in large buildings the coaxial cable losses add up quickly (4 dB/100 feet), resulting in signal losses. The base-station amplifier power requirements increase and the uplink receive sensitivity is degraded. Mobile units would therefore need to operate at higher power levels with the consequential reduction in talk time and battery change. In Figure 1.39 we see an active coaxial system using by directional amplifiers to compensate for cable RF losses.

In active RF coaxial systems, one of the most important endeavors is the site-specific engineering required for each installation to determine the proper location of bidirectional amplifiers in the coaxial layout. Amplifiers' placement can affect

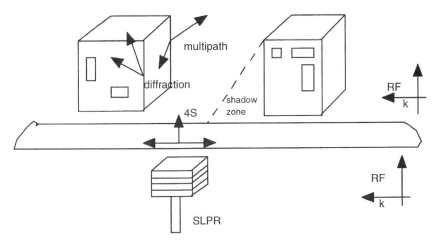

Figure 1.38   Example of the usage of an SLPR. (Modified from Ref. [6a] with permission from *Wireless Systems Design.*)

the system's noise and distortion measurements. In ideal environments amplifiers should be placed as close as possible to the antennas, though that is not always possible. A single amplifier can often support several antennas in the system through the use of RF splitters. Losses of up to 20 dB can often be observed in active coaxial systems before the signal is again amplified, causing the receiver sensitivity to be degraded. Finally, coaxial active systems are not easily upgradable. If coverage needs to be increased, it cannot be done by adding more antennas or cabling, since the noise figure can decrease and noise can be added inadvertently to the system.

Fiber-optic RF distribution uses small fiber-optic antennas mounted in a ceiling in order to provide coverage to a sector of a building. These antennas are connected by optical fibers to an RF distribution center, which provides the interface to the cellular system through connection to either a base station or a repeater. (See Figure 1.40.)

The fiber-optic antennas placed throughout the building will provide uniform coverage. Because fiber attenuation is very small, the RF distribution center and cellular equipment can be placed anywhere in the building. The signal losses in an optical system are very minute compared to those of coaxial systems. In the fiber-optic distribution system, cellular signals are split and sent to several fiber-optic transceivers to convert RF signals to optical signals. These optical signal cables connect to fiber-optic antennas. At the antenna, a photodiode converts the optical signals back to RF signals, which are amplified and radiated using a

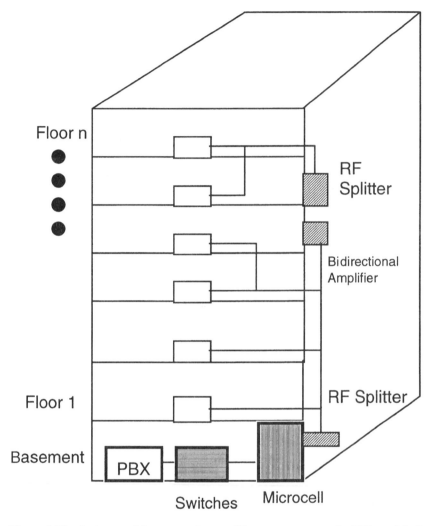

Figure 1.39   Active coaxial system using amplifiers to compensate for RF loss. (Modified from Ref. [6b] with permission from *Microwaves & RF*.)

suitable antenna. In the uplink mode, signals are received and amplified before a laser diode converts these RF signals to optical signals, for transmission to the RF distribution center. There is amplification of the input signal right after it is received at the fiber-optic unit, which makes this choice very suitable. At the RF distribution center, optical signals are converted to RF signals and combine with the signals from other fiber-optic antennas. All fiber-optic antennas that are

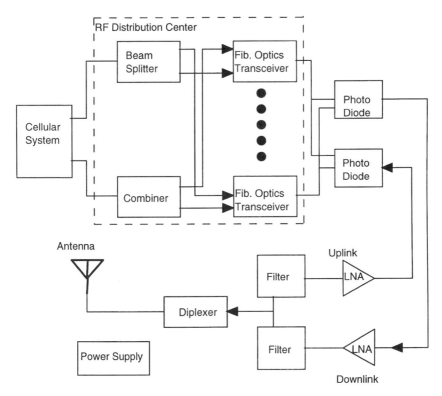

Figure 1.40   Fiber-optics RF distribution center for cellular system. (Modified from Ref. [6b] with permission from *Microwaves & RF.*)

served off the same splitter/combiner pair appear to the base station as if they were a single antenna. It is also possible to have multiple splitter/combiner pairs to configure a building as multiple cells.

In a fiber-optics system, the number of antennas per cell and the number of cells per building depends on the capacity and coverage requirements. In order to increase coverage, fiber-optic antennas or radio transceivers are added to the RF distribution center.

1.10   Use of Smart Antennas

Smart antennas will be used in the near future to increase the coverage and capacity (measured user/km^2) as well as signal quality (the ratio of signal to interference plus noise). There are basically two kinds of smart antennas:

switched-beam and adaptive-array. Switched-beam systems use a beam-forming circuit to form multiple fixed beams and a computer controller to choose the best beam among several options based on such performance criteria as received signal strength in analog systems and bit-error rate in digital systems. Switched beams are simple; they require little digital signal processing. The antennas used in switched systems are phased arrays. In a phased array, the phases of the exciting currents on each antenna (patch or dipole antennas) are scanned in a pattern to provide a maximum in a desired direction.

Simple switched-beam antenna systems are the simplest form of smart antennas, since in such systems a fixed-phase feed network (also known as a beam former) provides several beams with fixed directions with only one beam selected for the downlink and uplink. In adaptive systems, an adaptive array controls its pattern by the use of feedback to vary the phase weighting and/or amplitude weighting of the signals received by each element in order to obtain the received signal. Adaptive arrays have the capability of better rejecting interfering signals. Adaptive beam forming is easier to implement on the uplink than on the downlink if the frequencies are different. Smart antennas are capable of providing higher gain than omnidirectional or sector antennas, as shown in Figure 1.41. This increases the transmit and received range and hence the coverage area. Systems which use smart antennas can transmit at a lower power.

Another important capability of adaptive antennas is that of interference reduction. Furthermore, co-channel interference can be further reduced by having the base station steer the beams directionally toward the mobile unit. Interference with co-channel mobiles can only happen if both units are within the narrow

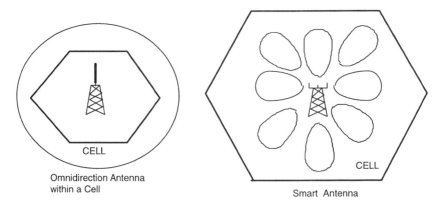

Figure 1.41 Comparison of an omnidirectional and a smart antenna.

widths of the directional beams. Therefore, interference rejection can be obtained by using directional beams. Directional beams on the downlink decrease the likelihood that the base station will interfere with co-channel mobile units, as shown in Figure 1.42.

The number of cells can then be reduced, increasing the spectral efficiency and capacity. To successfully increase capacity by reducing the number of cells, the system must therefore be able to detect unacceptable levels of interference. The base station can be configured to provide only those channels on which no interference is detected.

Smart antennas are very useful in CDMA. In a CDMA system all users occupy the same bandwidth and operate in an environment of constant co-channel interference. Each user is assigned a different spreading code, with the codes having low cross-correlation. The CDMA system will then assign a base-station receiver to each mobile user. The receiver correlates the incoming signal with the user's spreading code. In this approach the CDMA system can distinguish and receive signals from each mobile unit even in the presence of noise. The use of smart antennas can produce interference and increase capacity by allowing

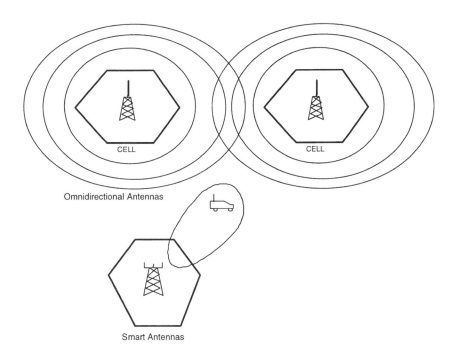

Figure 1.42   Smart antennas decrease chances for co-channel interference.

nonusers to share the same frequency within a cell. In Figure 1.43 the mobile unit of interest is represented by a mobile car unit, with other mobile units also present which are interpreted as interferers by the base station. A smart antenna contributes significantly so that other mobile units will not fall within the beamwidth of the antenna.

Among the multiple access systems such as TDMA, FDMA, and CDMA, the CDMA system is the most capable of taking advantage of smart antennas. CDMA systems do not need any special changes. Applications that may require interference reduction and interference rejection are very suitable to use with smart antennas. Adaptive beam-forming systems were originally used to avoid jamming in radars and military communications systems. Adaptive beam-forming uses algorithms that interactively adjust the weighting of the signals, and this approach results in deep pattern nulls in the direction of interference. Adaptive beam-forming can be implemented by weighting and combining the element signals

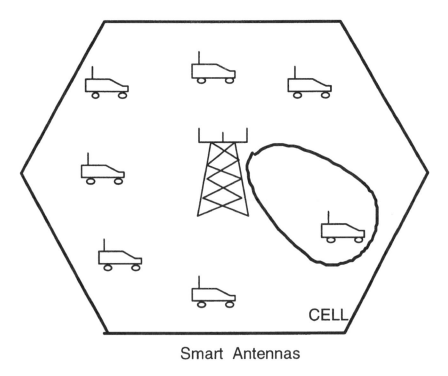

Smart Antennas

Figure 1.43 Smart antennas allow selectivity to avoid interferers.

of an intermediate frequency (IF) or at baseband. In the former approach, signals from each array element are weighted and combined at IF. These signals are converted to baseband for demodulation as shown in Figure 1.44.

1.10.1   PROPAGATION MODELS FOR SIMULATING INTERFERENCE

The physical medium between antennas where electromagnetic waves are propagated is called the propagation channel. Even the different kinds of obstacles which would influence the propagation of electromagnetic waves, whether they are static or time-varying obstructions, are also considered part of the propagation channel. In certain applications when the channel variations are quite slow compared to the transmission rate, such channels are called quasi-static.

Regardless of the type of wireless system, the same physical principles govern the propagation of radio waves. The known mechanisms of reflection, diffraction,

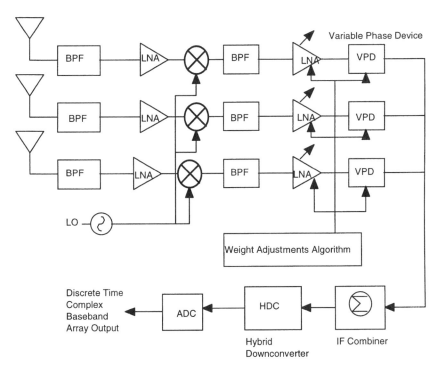

Figure 1.44   Adaptive beam-forming method for creating nulls in the direction of interference.

and scattering can distort the transmitter signal. In order to make a good evaluation, it is necessary to understand such mechanisms.

1.10.2 MULTIPATH INTERFERENCE

In a typical land mobile system such as cellular phones and digital PCS, there are multiple propagation paths that are reflected for every direct path on a line of sight. Actually, in most cases there is really no complete direct line-of-sight propagation between the base station antenna and the mobile antennas because of many reflection scattering paths as shown in Figure 1.45.

In such an environment, the propagation path varies randomly, and because of the multiplicity of such paths, the term multipath propagation has been used. Even the slowest movement can cause time-variable multipath and therefore random reception of time-varying signals. Radio propagation in such environments is exposed to three basic kinds of interference: multipath fading, shadowing, and path loss. Multipath fading is of three kinds: envelope fading (nonfrequency-selective amplitude distribution); Doppler spread (time-selective phase noise); and time-delay spread, which is the variable propagation distance of reflected signals causing time variations in the reflected signals. (See Figure 1.46.)

1.10.3 ENVELOPE FADING BASICS

We first have to assume that the base station is transmitting a constant envelope phase modulated signal $S_B(t)$ given by

$$S_B(t) = A e^{j(\omega_c t + \phi_n(t))}, \tag{1.21}$$

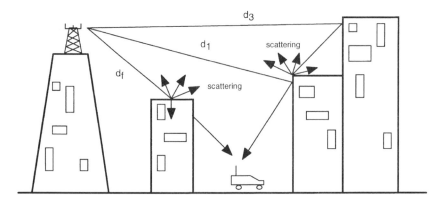

Figure 1.45 Illustration of the true lack of direct path on a line of sight.

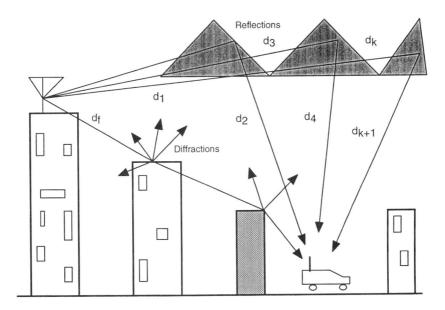

Figure 1.46 A broad view of multipath interference.

where A is a magnitude constant, ω_c is the angular carrier frequency, and $\phi_n(t)$ is the phase or information-bearing signal known as baseband. The time-variable random propagation medium $m(t)$ is represented as

$$P(t) = r(t)e^{j\phi_r(t)}, \tag{1.22}$$

where $r(t)$ is the time-variable envelope and $\phi_r(t)$ is the time-variable random phase of the propagation medium. The envelope of the random propagation medium $r(t)$ is separated into two terms: average or long-term fading $m(t)$, and short-term or fast multipath fading $r_o(t)$, where $r_o(t)$ has unit mean value. Therefore,

$$P(t) = m(t)r_o(t)e^{j\phi_r(t)}. \tag{1.23}$$

The constant envelope transmitted signal $S_B(t)$ is being multiplied by the transfer function of the propagation medium $P(t)$. The received signal at the mobile unit is given by

$$S_m(t) = S_B(t)\,P(t) = Am(t)r_o(t)e^{j(\omega t + \phi_n(t) + \phi_r(t))}. \tag{1.24}$$

Notice from this expression that a time-variable random phase modulation component $\phi_r(t)$ has been introduced by the mobile main propagation medium. The

$\phi_r(t)$ random phase variation is the main cause of frequency spreading known as Doppler spread.

It has been theoretically shown [7] that the received signal envelope has a Rayleigh distribution when the number of incident plane waves propagating randomly from different directions is large enough and when there is no predominant line-of-sight component. The Rayleigh distribution is the most frequently used distribution in mobile channels.

1.10.4 DOPPLER SPREAD BASICS

In the previous paragraphs it has been shown that the $\phi_r(t)$ phase change is related to the rate of change of the fast-falling component $r_0(t)$. This phase variation includes FM noise on the carrier being received. In the work of Jakes [7], it was demonstrated that the baseband spectrum of the random FM noise is approximately twice the maximum Doppler spread. The maximum Doppler frequency is given by

$$f_d = vf/c, \qquad (1.25)$$

where v is the speed of the mobile, including the speed of the mobile environment (m/sec), f is the radio frequency (Hz), and $c = 3.0 \times 10^8$ m/sec.

The Doppler spread is the spectral width of a received carrier when a single sinusoidal carrier is transmitted through the multipath channel; because of Doppler spread we receive a distorted signal spectrum with spectral components between $f_c - f_d$ and $f_c + f_d$. Finally, coherent time C_T is defined as the required internal time to obtain an envelope correlation of 0.9 or less. It is inversely proportional to the maximum Doppler frequency and defined by

$$C_T = 1/f_d.$$

1.10.5 TIME DELAY SPREAD BASICS

The fundamentals of delay spread are illustrated in Figure 1.47. At $t = 0$, the first burst of the base station is transmitted from the base station to the mobile unit. The direct line-of-sight path has a length of d_1(m) and a propagation delay of t_0. On the other hand, the reflected multipath signals along

$$d_1 + d_2, d_3 + d_4, \ldots, d_k + d_{k+1}$$

have a much larger propagation delay that the direct line of sight arriving along the d_f path and often have magnitude comparable to the direct line of sight. The signal received by the mobile through reflected path $d_1 + d_2$ has a propagation

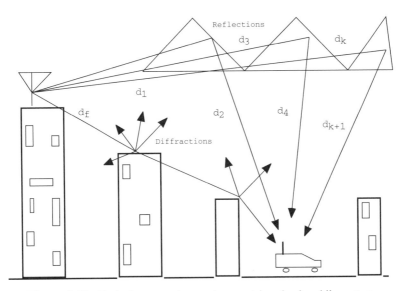

Figure 1.47 Typical propagation environment in a land mobile system.

delay of $t_1 + t_2$, and that through the reflected path $d_3 + d_4$ has a propagation delay of $t_3 + t_4$. The delay of the path $d_1 + d_2$ relative to the direct line-of-sight signal is given by $t_r = t_1 + t_2 - t_f$, which is called the first arrival delay. In the same manner, the second arrival delay is given by $t_{r2} = t_3 + t_4 - t_f$. In a realistic multipath interference environment, a large number of delayed components are added, forming a power delay profile. The extent of the power delay profile is what is called the delay spread. The power delay profile is really a density function given by

$$P(t) = \frac{S(t)}{\int S(t)dt}, \tag{1.26}$$

where $S(t)$ is the measured power delay profile as shown in Figure 1.48. The average delay measure with respect to the first arrival delay is given by

$$t_e = \int (t - t_{R1}) P(t) \, dt. \tag{1.27}$$

Finally, the rms delay term is a measure of delay spread. It is the standard deviation about the mean excess delay and is given by

$$t_{rms} = \sqrt{\int (t - t_e - t_{R1})^2 P(t) \, dt}. \tag{1.28}$$

A simpler worst-case delay was introduced by Feher [8]. The maximum delay spread, abbreviated as t_{max}, is calculated based on basic system parameters such

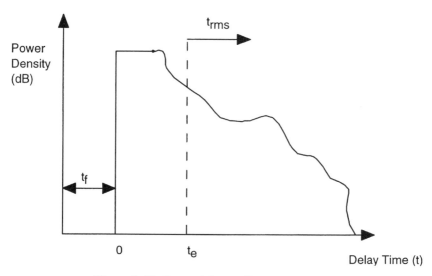

Figure 1.48 Power delay profile or delay spread.

as transmitted power (P_t), received power at threshold (P_{rmin}), and radio frequency f_c. It is assumed that the direct line-of-sight path is short, and that afterwards scattering and severe signal attenuation occur in this direct path. Feher's bound on the roundtrip propagation delay is given by

$$t_{max} = \frac{d_{max}}{c},$$

where

$$d_{max} = \left[\frac{P_t G_t G_R (\lambda/4\pi)^2}{P_{rmin}} \right]^{1/2}. \tag{1.29}$$

P_t is the transmitted power, G_t and G_R are the transmit and receive antenna gains, f is the carrier frequency, $\lambda = c/f$, $P_{rmin} = KTB$ (Hz)$F + C/N$, B(Hz) is the receiver bandwidth, F is the noise figure of the receiver, and C/N is the required carrier-to-noise ratio in the receiver bandwidth.

1.11   Path Loss

Path loss is the average value of log-normal shadowing, which is often called coverage area. Shadowing is mainly caused by terrain features in mobile propagation environments. It imposes a slowly changing average on the Rayleigh fading statistics. Even though there is no mathematical model for shadowing, a log distribution with a standard deviation of 5 to 12 dB has been found to best fit the experimental data in a typical urban area.

Most of the land mobile systems and PCS work in a non-direct-line-of-sight environment. Based on empirical data, a general model was developed for non-line-of-sight propagation and is given by

$$L(d) \propto L_B \left(\frac{d}{d_o}\right)^{-n}, \tag{1.30}$$

where n is the path loss exponent, typically in the range of $3.5 \le n \le 5$ (outdoor) and $2 \le n \le 4$ (indoor); d is the separation between transmit and receive antennas; d_o is the reference distance of free space propagation corner distance; $L_B(\text{dB}) = 27.56 - 20 \log f$ (MHz) $- 20 \log R$ (m); and L is the propagation loss of the combined non-line-of-sight and line-of-sight signal path.

The exponent n indicates how fast the loss increases with distance. The reference distance d_o assumes that we have free space propagation between the antenna and d_o.

There have been empirical models used in the past to predict the average path loss along a given path, especially from the base station antenna to the antenna in the mobile unit. Extensive measurements performed by Okomura have led to an empirical formula for the medium path loss L_p (dB) between two isotropic antennas. This formula is given by

$$L_p = \begin{cases} A + B \log (R) & \text{for urban area} \\ A + B \log R - C & \text{for suburban area} \\ A + B \log R - D & \text{for open area} \end{cases}, \tag{1.31}$$

where R is in kilometers, the radio carrier frequency is f_c (MHz), the base station height is h_b (m), and the mobile station antenna height is h_m (m). The values of A, B, and C are given by

$$A(f_c, h_b, h_m) = 69.55 + 26.16 \log(f_c) - 13.82 \log(h_b) - a(h_m)$$
$$B(h_b) = 44.9 - 6.55 \log(h_b)$$
$$C(f_c) = 2\left[\log\left(\frac{f_c}{28}\right)\right]^2 + 5.4$$
$$D(f_c) = 4.78 \log^2(f_c) - 19.33 \log(f_c) + 40.94, \tag{1.32}$$

where

$$a(h_{\rm m})$$

$$= \begin{cases} [1.1 \log(f_c) - 0.7]h_{\rm m} - [1.56(f_c) - 0.8] & \text{for medium or small city} \\ 8.28[\log(1.54\, h_{\rm m})]^2 - 1.1 & \text{for } f_c \geq 200\,\text{MHz} \\ 3.2[\log(11.75\, h_{\rm m})]^2 - 4.97 & \text{for } f_c \geq 400\,\text{MHz for large city} \end{cases}.$$

This equation can be used if these conditions are satisfied:

f_c = 150 to 1500 Mhz $h_{\rm m}$ = 1 to 10 m
$h_{\rm b}$ = 30 to 200 m R = 1 to 20 km.

Another interesting term, this one given by Feher [8], is that of communication range or the maximum distance that can be covered for free space loss line-of-sight propagation conditions, which is given by the expression

$$d_{\max} = \left[\frac{P_t G_t G_R \left(\dfrac{\lambda}{4\pi d_{\rm o}}\right)^2}{P_R} \right]^{1/n} d_{\rm o}, \tag{1.33}$$

where $d_{\rm o}$ is the radio-wave propagation line of sight.

1.12   Co-channel Interference

Co-channel interference occurs when two or more independent signals are transmitted simultaneously by the same frequency band. The same frequencies reuse co-channel interference. This can be observed in Figure 1.49. In the figure the frequencies f_1 through f_7 are reused in the $k = 7$ system. If the mobile unit is at location M6, it will receive the desired signal on frequency f_6 from the nearest base station 6C. In the same manner the mobile unit receives, in the same frequency band, independent interference signals from base stations 6A and 6B. The ratio of desired average carrier power S from the nearby base station 6C to the average interference power from the distance base stations 6A and 6B is given by

$$a_T = \frac{D_{\rm a}}{R} + \frac{D_{\rm b}}{R}, \tag{1.34}$$

where $D_{\rm a}$ is the distance from base station 6A that transmits the same frequency as 6C, $D_{\rm b}$ is the distance from base station 6B that transmits the same frequency,

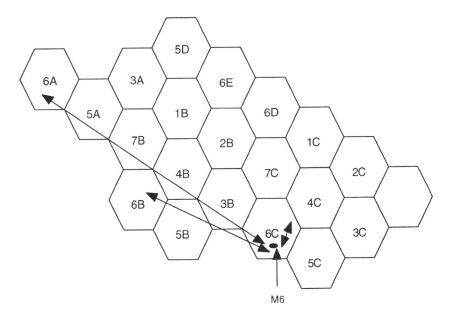

Figure 1.49 Illustration of co-channel interference.

and R is the coverage radius of the base-station transmitter of one cell. The relation between D and R for hexagonal cells sharing k frequencies is

$$D = \sqrt{3K}\,R.$$

The co-channel-interference-caused S/I ratio received at a desired base station from a given number M of interferers (i) is

$$\text{Co-channel interference} = \frac{S}{N_T + \sum_{i=1}^{M} I_i}, \tag{1.35}$$

where N_T is the total noise power in the receiver bandwidth and is given by

$$N_T = KT\,\Delta f\, F. \tag{1.36}$$

K is the Boltzmann constant (-228.6 dBW/K), T is the absolute temperature in kelvins, Δf is the double sideband noise bandwidth of the receiver, and F is the noise figure of the receiver.

1.13   Adjacent Channel Interference

Adjacent channel interference can be caused by modulation and nonlinearities within electronic components. In such cases the transmitted signal is not really band limited; instead, radiated power will go into adjacent channels. It is assumed that several adjacent channels are causing interference with each other as shown in Figure 1.50:

$$\Delta f_1 = f_{c2} - f_{c1}$$
$$\Delta f_2 = f_{c3} - f_{c1}.$$

The interference power is caused by the first upper and lower interfering signals.

In the U.S. cellular system (IS-54), the first adjacent channel interference is specified at -26 dB below the desired carrier power; W_r ("brick wall" receiver channel) = 30 kHz. In the digital European cordless telephone system, the channel spacing is about $\Delta f = 18$ MHz, whereas the adjacent channel interference is specified as $W_r = 1.1$ MHz.

Feher [8] proposed the following expression for adjacent channel interference:

$$\text{Adjacent channel interference} = \frac{\int_{-\infty}^{\infty} G(f)|H(f - \Delta f)|^2 \, df}{\int_{-\infty}^{\infty} G(f)|H(f)|^2 \, df}, \quad (1.37)$$

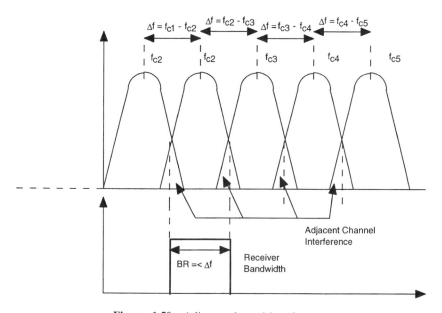

Figure 1.50   Adjacent channel interference.

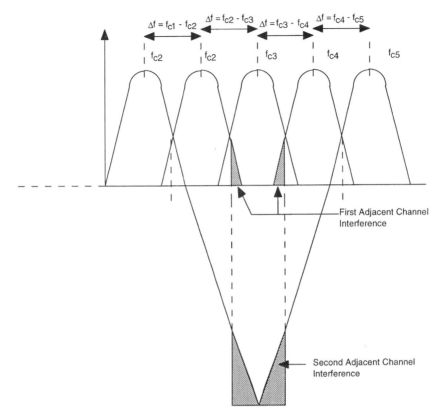

Figure 1.51 First and second adjacent channel interference. (*Wireless Digital Communications: Modulation and Spread Applications* by Feher, © 1995. Adapted by permission of Prentice-Hall, Inc., Upper Saddle River, NJ.)

where $G(f)$ is the power spectral density of the signal, $H(f)$ is the receiver bandpass filter transfer function, and Δf is the carrier spacing between adjacent channels.

1.14 Rayleigh Fading in Quasi-static Systems

A Rayleigh-faded system is a slow fading channel if

$$f_{\text{Doppler}} \, T_{\text{b}} < 10^{-4},$$

where f_{Doppler} is the Doppler spread and $T_{\text{b}} = 1/f_{\text{b}}$, f_{b} being the bit rate. In these slowly faded channels, Doppler shift does not have a degrading impact on the

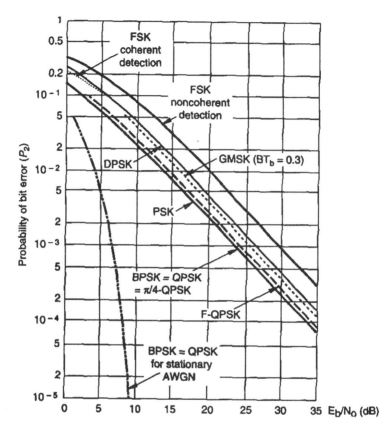

Figure 1.52 Probability bit error performance of linearly amplified coherent BPSK, QPSK, π/4-QPSK, and wideband FSK and of noncoherent DPSK and FSK in a slow Rayleigh-faded system without delay spread. Nonlinearly amplified coherent GMSK and FQPSK are also illustrated. (*Wireless Digital Communications: Modulation and Spread Applications* by Feher, © 1995. Adapted by permission of Prentice-Hall, Inc., Upper Saddle River, NJ.)

probability of bit error P_e, which is a function of (E_b/N), where E_b is the average energy of a modulated bit and N is the noise spectral density at the demodulator input. The probability of bit error P_e as a function of E_b/N for ideal BPSK, FSK, QFSK, and π/4-QPSK coherent systems and noncoherent DPSK and noncoherent, wideband large-deviation-index FSK systems is illustrated in Figure 1.52. The probability of symbol error P_s as a function of E_b/N of several coherent and differentially demodulated PSK systems is predicted in the figure [8].

Chapter 2 | Space Environment Effects in Communications

2.0 Natural Radio Noise Environments

We provide material useful to the communications engineer, as well as a mechanism for entry into the literature on the subject, across the radiofrequency band up to 300 GHz. This is done by first discussing some basic concepts, then considering noise as it appears at the surface of the earth. As frequency is increased, one passes from a regime dominated by noise due to lightning to one dominated by extraterrestrial noise, and then to the regime where thermal atmospheric emissions are controlling. A few miscellaneous items such as quantum noise and the effects of the ground on antenna temperature are then considered. Noise at elevated locations is discussed briefly in Section 2.13 for the case of earth satellites, particularly where downward-looking antennas are involved.

The terminology used in this work will conform, as far as possible, with the following definitions:

Noise factor is the ratio of noise power measured at the output of the receiver to the noise power which would be present at the output if the thermal noise due to the resistive component of the source impedance were the only source of noise in the system; both noise powers are determined at an absolute temperature of the source equal to $T = 293$ K.

Noise temperature is the value by which the temperature of the resistive component of the source impedance should be increased, if it were the only source of noise in the system, to cause the noise power at the output of the receiver to be the same as in the real system.

Width of the effective overall noise band is the width of a rectangular frequency response curve, having a height equal to the maximum height of the receiver response curve and corresponding to the same total noise power.

2.1 Noise Factor

The overall operating noise factor, f, of a receiving system is given by

$$f = f_a + (l_c - 1)\left(\frac{T_c}{T_0} + l_c\right)(l_t - 1)\left(\frac{T_c}{T_0} + l_c\right)l_t\,(f_r - 1). \qquad (2.1)$$

f_a is the external noise factor, defined as

$$f_a = \frac{P_n}{k\,T_0\,b}.$$ (2.2)

F_a is the external noise figure, defined as

$$F_a = 10\,\log f_a.$$

P_n is the available noise power in watts from a lossless antenna (the terminals of this lossless antenna do not exist physically, but this is the only appropriate point in the receiving system to reference the various noise factors).
k is Boltzmann's constant, 1.3802×10^{-23} joule/K.
T_0 is the reference temperature in kelvins, taken as 288 K.
b is the noise power bandwidth of the receiving system in hertz.
l_c is the antenna circuit loss (available input power to available output power).
T_c is the actual temperature, in kelvins, of the antenna and nearby ground.
l_t is the transmission line loss (available input power/available output power).
f_r is the noise factor of the receiver ($F_r = 10\,\log f_r$: noise figure in decibels).

Relation (2.2) can be written

$$P_n = F + B - 204 \text{ dBW,}$$ (2.3)

where $B = 10\,\log b$, $P_n = 10\,\log P_n$, and $10\,\log kT_0 = -204$ dB.

2.2 Field Parameters

In order to relate P_n and F_a to the electric field strength and noise power flux density, we introduce the effective aperture of an isotropic radiator A_i:

$$A_i = \frac{\lambda^2}{4\pi}.$$ (2.4)

An arbitrary antenna of 100% efficiency and gain g relative to an isotropic radiator then has a capture area

$$A = g,A_i = g\frac{\lambda^2}{4\pi}.$$ (2.5)

The available noise power, P_n, is related to the noise power flux density, S, by

$$P_n = S,A = Sg\frac{\lambda^2}{4\pi}.$$ (2.6)

The field strength, E, is related to the power flux density by

$$E^2 = 120\pi S, \tag{2.7}$$

or

$$E[dB(1 \ \mu V/m)] = S \ dB(W/m^2) + 145.8, \tag{2.8}$$

where $E[dB(1 \ \mu V/m)] = 20 \log E(V/m) + 120$, and $S[dB(W/m^2)] = 10 \log S$ (W/m^2).

Combining (2.2) through (2.8), we get

$$E(dBu) = 20 \log f_{MHz} + F_a + B + G - 96.8, \tag{2.9}$$

where $dBu = dB(1 \ \mu V/m)$ and $G = 10 \log g$ is the gain in decibels of the antenna over an isotropic radiator.

There is an implicit assumption in relations (2.5)–(2.9) that the noise may be represented by a plane wave with a definable direction of propagation. In fact, the ambient noise may be decomposed into a spectrum of plane waves.

2.3   Antenna Temperature

The concept of antenna temperature T_a is analogous to that for f_a in Eq. (2.2). $4R \ kT_b$ is the mean square noise voltage in a resistor of R ohms; it follows that the noise power P_n transferred to a matched load is

$$P_n = \frac{4R \ kTb}{R} \cdot \left(\frac{R}{2R}\right) = kTb, \tag{2.10}$$

where T is the temperature of the resistor in kelvins and b is the bandwidth in hertz. Unfortunately, integrating P_n in Eq. (2.10) from zero to infinity in frequency results in an infinite power. The remedy is given by the application of quantum mechanics to equipartition theory, which yields

$$P_n(f) = \frac{hf}{\exp(hf/kT) - 1} \quad (W/Hz)$$

$$P_n(f) = kT \quad \text{for} \quad hf << kT, \tag{2.11}$$

where h is Planck's constant ($h = 6.625 \times 10^{-34}$ joule-sec). At room temperature (288 K), $kT = hf$ occurs at 6×10^{12} Hz or 6 THz, so this high-frequency complication can be ignored at radio frequencies. However, if the resistor is cooled to 3 K, the value at which $kT = hf$ decreases to 60 GHz and quantum noise becomes a consideration.

Let us now rewrite the expression for external noise factor in Eq. (2.2) in light of our definition for noise temperature:

$$f_a = \frac{T_a}{T_0} = \frac{P_n}{kT_0 b}$$

or   (2.12)

$$T_a = \frac{P_n}{kb}.$$

T_a is antenna temperature due to external noise measured in kelvins. Note that T_a is linearly related to P_n, the available noise power.

2.4 Brightness Temperature

The terms antenna temperature, sky noise temperature, and brightness temperature are frequently, and at times quite properly, used interchangeably. In this chapter we will make the following distinctions:

Antenna temperature T_a includes contributions from the sidelobes as well as the main lobe of the antenna, and hence usually includes a contribution from the surface of the earth.

Sky noise temperature T_{sky} does not include emission from the ground into the sidelobes of the antenna, but does include atmospheric emission into the antenna sidelobes.

Brightness temperature T_b will be used to describe noise coming from a particular direction, as if seen with an antenna of infinitely narrow beamwidth.

2.5 Thermal Radiation

Energy is radiated by a blackbody at temperature T and frequency f in accordance with Planck's radiation law,

$$b = \frac{2hf^3}{c^2} \frac{1}{\exp(hf/kT) - 1},$$ (2.13)

where

b = brightness (W/m^{-2} Hz^{-1} rad^{-2})
h = Planck's constant = 6.63×10^{-34} joule sec
k = Boltzmann's constant = 1.38×10^{-23} joule/K
f = frequency (Hz)
T = temperature of the blackbody (K).

The radio frequencies fall to the left of the peak for all realizable temperatures (for 3 K the peak occurs at 0.97 mm), so blackbody or thermal radiation at radio frequencies may be approximated by the Rayleigh–Jeans law,

$$b = \frac{2kT}{\lambda^2} = 22kT(f_{\text{GHz}})^2. \tag{2.14}$$

Note that for a given temperature the radiated energy (brightness) is proportional to the square of the frequency. This is different from the case for a resistor, where frequency is not a factor in the noise voltage in the radio range. The term brightness temperature derives from the term ''brightness'' as used for radiant energy in Eqs. (2.13) and (2.14).

Thus, from Eq. (2.14) we obtain

$$T_{\text{b}} = \frac{b\lambda^2}{2k} = \frac{b}{22k(f_{\text{GHz}})^2}, \tag{2.15}$$

where T_{b} is the brightness temperature of a thermal source in kelvins, b is the brightness (W/m^{-2} Hz^{-1} rad^{-2}), λ is the wavelength in meters, k is Boltzmann's constant = 1.38×10^{-23} J/K, and f is frequency (GHz).

2.6 Natural Noise at the Earth's Surface

Natural noise at the earth's surface is a generally decreasing function of frequency as is illustrated in Figure 2.1. Below about 5 Hz noise is predominantly geomagnetic micropulsations. From 5 Hz to 10 MHz, the dominant source is lightning, and the received noise goes by the name atmospherics, or simply spherics. The ionosphere plays a strong role in the propagation of noise from distant lightning strokes, with the result that significant temporal variations exist. From 10 to 100 MHz, natural noise is a combination of atmospherics and extraterrestrial noise (galactic and solar). Extraterrestrial noise dominates between 100 and 1000 MHz (1 GHz), but by 10 GHz, emission from the atmosphere itself (neutral gases, clouds, and rain) has taken over and becomes increasingly important as frequency

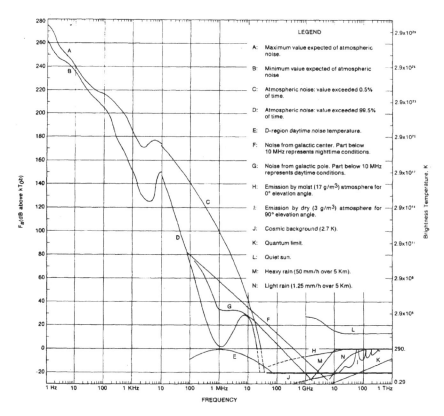

Figure 2.1   Overview of natural noise levels in the radio spectrum (after Spaulding and Hagn [12]).

increases to the limit of the radio spectrum (arbitrarily taken to be 300 GHz). The decade 1 to 10 GHz is the quietest from the natural noise standpoint and is understandably popular for sophisticated communication systems involving long line-of-sight paths, such as those employed by the earth space services.

2.7   Atmospherics

Noise due to lightning is the most thoroughly studied and best-understood type of natural noise and for that reason will receive short shrift in this chapter. Worldwide maps of natural noise for frequencies below 30 MHz were first prepared in the United States and the U.K. in World War II. The next major compilation appears as CCIR Report 322 [10] and provides data for frequencies of 10 kHz to 100 MHz based on maps at 1 MHz. In his book *VLF Radio Engineering* [9], A. D. Watt provides world maps at 10 kHz for use where the primary interest is in very low frequencies. Report 322 has been criticized for its treatment of noise in equatorial climates (e.g., CCIR Report 258-3) and there has been considerable work and an extensive literature developed on this aspect.

2.8 Extraterrestrial Noise

Extraterrestrial noise is the subject of radio astronomy. Radio astronomers are interested in the information content in the noise, and a formidable literature exists. The noise sources to be considered include the following:

- The sun
- The galaxy
- The cosmic background
- Discrete stellar sources
- The moon and planets

A good review of these noise sources for telecommunications use has been given by Boischot [11].

The emission from the sun at radio frequencies is a complex phenomenon with a well-developed literature [13–16]. Following Maxwell [17], this emission consists of the following:

1. A background thermal emission, which results from free transitions of electrons in the field of ions. The thermal emission processes are fairly well understood. Present radio models for the background radiation are in good accord with the observed data.

2. A slowly varying component which is related to the total area of sunspots visible on the sun and is most prominent at centimeter and decimeter wavelengths. This component is difficult to distinguish at frequencies lower than 200 MHz. It is believed to be thermal in origin.

3. Transient disturbances, sometimes of great intensity, which originate in localized active areas. These bursts are most intense at meter and decimeter wavelengths, but are visible at times throughout the spectrum. These bursts are classified by their spectral behavior as:

- Type I noise storm bursts
- Type II slow drift bursts
- Type III fast drift bursts
- Type IV continuum bursts

The quiet sun's radiation has a brightness temperature of 6000 K at frequencies above 30 GHz, which corresponds to the physical temperature of the photosphere from which it emanates. The brightness temperature at frequencies below 30 GHz exceeds that of the photosphere because the emission is taking place at higher altitudes in the solar atmosphere (chromosphere and corona) where the temperature can exceed 1 million K.

Galactic noise is not thought to be thermal in nature as the frequency dependence is wrong. According to the Rayleigh–Jean law the brightness (flux density), b, of a discrete object of uniform temperature is

$$b = \frac{2kT}{\lambda^2} \, \Omega \qquad (2.16)$$

where Ω is the source solid angle. Let the variation of brightness b with wavelength be expressed by the proportionalities [20]

$$b \propto \lambda^n, \; T_b \propto \lambda^{2+n}, \quad \text{or} \quad T_b \propto f^{-(2+n)},$$

where n is the spectral index, a dimensionless quantity, and f is the frequency. By this definition a thermal source has a spectral index of $n = -2$, while a positive value of n represents a brightness temperature which decreases rapidly with frequency. Nonthermal sources have an average spectral index of about 0.75 [18], but the sky background may be represented by a shallower spectral index ($n = 0.2$ to 0.6) below about 250 MHz, and the higher value ($n = 0.6$ to 0.9) above 250 MHz. Figure 2.2 is a mass plot of galactic noise measurements from 10 MHz to 1 GHz. Shown in this figure with solid horizontal lines are nighttime measurements made by Yates and Wielebinski [19] at five different frequencies (14 to 85 GHz) with scaled antennas aimed toward declination $\delta = 34°$. The brightness temperature extremes from the Ko and Kraus survey [20], given in Figure 2.3, are plotted with solid dots at 250 MHz. Two computer-derived maps published by R. E. Taylor [21] and reproduced in the ITT *Handbook* [22] appear at 136 MHz and 400 MHz in Figure 2.2. The dashed-line values are drawn between values of the galactic center and galactic pole given by Kraus [18].

Most surveys of brightness temperature at frequencies above 400 MHz are plotted in galactic coordinates and possess a wealth of detail (due to the smaller beamwidth employed) which makes a single-page-sized chart impossible. A

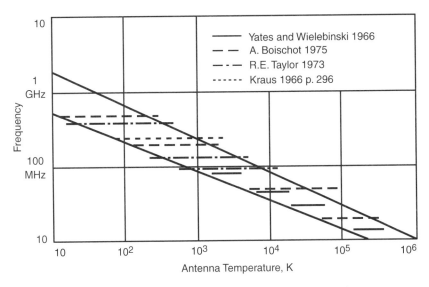

Figure 2.2 Observed and predicted galactic noise levels.

discussion of the major surveys and examples may be found, for example, in *Radio Astronomy and Cosmology* [23], Haslum *et al.* [24], and Berkhuijsen [25]. The survey shown in Figure 2.3 is from Ko and Kraus [20] and is reproduced in Kraus' book *Radio Astronomy* [18] and in CCIR Report 720-1 [26]. It is for 250 MHz. The brightness temperature in kelvins is obtained by multiplying the number on each curve by 6 and adding this product to 80. For example, the brightness temperature of the 7 contour is $T_b = 7 \times 6 + 80 = 122$ K. The map is in celestial coordinates and the values $\pm\ \theta_o$, the bounds of declination of the sky behind a geostationary satellite as viewed from any location on the earth, are shown as two horizontal lines. For example, an earth station at 40°N latitude viewing a geostationary satellite on its own longitude would scan a strip of sky behind the satellite defined by declination

$$\delta = -\tan^{-1} = \frac{\sin 40^0}{6.6 - \cos 40^0} = -6.3^0.$$

The maximum brightness temperature (right ascension = 18 h 40 m) at 250 MHz reaches 850 K in this survey. The corresponding brightness temperature for frequencies above 250 GHz could be estimated using a spectral index of 0.75, and for frequencies below using a spectral index of 0.4 from the proportionalities given earlier. An isotropic noise temperature contribution of 2.7 K known

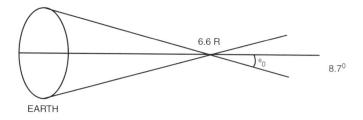

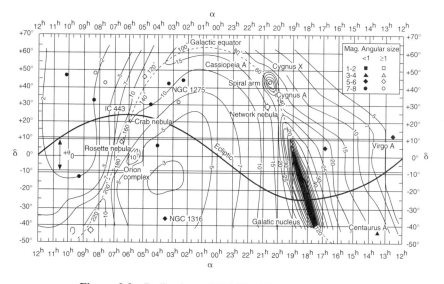

Figure 2.3   Radio sky at 250 MHz (after Ko and Kraus).

as cosmic noise (Penzias and Wilson [27]) can be observed at frequencies of a few gigahertz and higher. It is considered to be the residual radiation from events occurring during the origin of the universe.

The most prominent discrete stellar noise sources are plotted in celestial coordinates in Figure 2.4. Description of their origins, intensities, and locations are given by Boischot, Wielebinski, and Howard and Maran [11,28,29]. Strong noise sources and those useful for system calibration in the vicinity of the geostationary orbit are plotted in the figure from data provided in Wielebinski [28]. The moon and planets behave essentially as thermal sources, with the exception of Jupiter, where very interesting special effects take place [18].

Emission from the atmosphere is related to absorption by atmospheric constituents. It will be treated first for the clear atmosphere, then for fog and clouds,

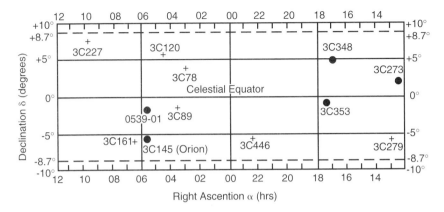

Figure 2.4 Strongest sources and calibration sources occurring between ±10°.

and finally for rain. By Kirchhoff's law, the emission from a gas in local thermodynamic equilibrium must equal its absorption, and this must apply at any frequency [30]. The brightness temperature in a given direction through the atmosphere is then given by radiative transfer theory [31],

$$T_b = \int_0^\infty T(l_i)\gamma(0,\omega)\, e^{-\tau(l_i,0)}\, dl + T_\infty\, e^{-\tau_\infty}, \tag{2.17}$$

where $T(l_i)$ is the local ambient temperature, $\gamma(0,\omega)$ is the local absorption coefficient taken for two atmospheric constituents, molecular oxygen and water vapor, and

$$\tau(l_i,0) = \int_0^{l_i} \gamma(0,\omega)\, dl' \tag{2.18}$$

is the optical depth between the emitting element and the receiver. $T_\infty \exp(-\tau_\infty)$ is the temperature from outside the atmosphere, in the direction in question, reduced by the optical depth T through the atmosphere in that direction. This extraterrestrial component, normally less than 3 K, is omitted in the plots. Brightness temperature from the gaseous atmosphere is given in Figure 2.5 [32,33]. The surface value of temperature (15°C) in the U.S. Standard Atmosphere will not support 17 g/m^3 of water vapor. Hence, a tropical model atmosphere for 16°N latitude with a higher surface temperature (29.4 K) has been substituted. The brightness temperate due to fog and clouds is more difficult to determine to the same degree of accuracy as for the clear atmosphere. An approach for estimating it is given here. Further details and computer analyses are given in

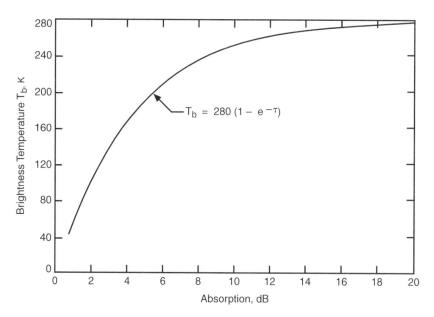

Figure 2.5 Brightness temperature due to atmospheric absorption under isotropic conditions.

Slobin [34]. The full radiative transfer expression for brightness temperature may be simplified by assuming an isothermal atmosphere. As $d\tau(l_i,0)$ $\gamma(0,\omega)$ dl, one may rewrite the previous equation for T_b as

$$T_b \approx T_0 (1 - e^{-\tau_\infty}) + T_\infty e^{-\tau_\infty}, \qquad (2.19)$$

corresponding to the first and second terms of Eq. (2.17), respectively. The second term, normally small, is discarded, yielding

$$T_b \approx 280 (1 - e^{-\tau}) \approx 280 (1 - 10^{-A/10}), \qquad (2.20)$$

where $\tau = A$ (dB)/4.34 is the attenuation through the atmosphere in the direction of interest and the arbitrary value $T_0 = 280$ K has been inserted to approximate that of an equivalent isothermal atmosphere. The effect of Eqs. (2.18) and (2.19) is to relate brightness temperature monotonically to absorption of a radio wave transiting the atmosphere from the same direction (see Figure 2.5). This absorption may be due to the gaseous atmosphere or to droplets of liquid water (fog, cloud, or rain). However, to make use of expression (2.20) and Figure 2.5 it is necessary to obtain the total absorption due to the atmosphere, that is, clear air plus liquid water.

Most fog and cloud droplets are less than 20 microns (0.02 mm) and almost all are less than 60 microns (0.06 mm) in diameter [33,34]. While both can be ice as well as water, the liquid water case yields much higher absorption than ice and will be the one considered here. Attenuation of radio waves by fog and cloud can be calculated by Rayleigh scattering theory [35] if the droplets are small compared to the wavelength (true up to 1 THz where λ = 300 microns). The power attenuation constant is then proportional to the liquid water content ρ_l, shown by

$$\alpha_p = k_1 \rho_l \ (dB/km).$$

The coefficient k_1 is a function of frequency and temperature and is given in Figure 2.6. The missing ingredient in most instances is the integrated water

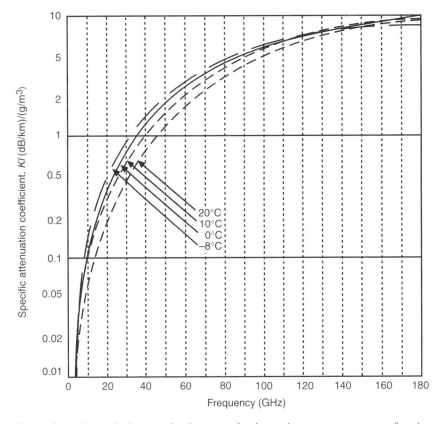

Figure 2.6   Theoretical attenuation by water cloud at various temperatures as a function of frequency.

content through the fog or cloud. Most clouds [34] will exhibit densities of less than 0.5 g/m³ of liquid water, the exception being cumulonimbus and nimbostratus clouds. Even the latter two rarely exceed densities of 1 g/m³ of liquid water. A rough estimate can be made using Figure 2.6 and assuming a cloud thickness and density. To this needs to be added a gaseous absorption contribution. This may be estimated from the values of zenith absorption given in the literature [36] or from Figure 2.7,

$$A_1 = k_1 \rho_1 l + A_0 \csc \theta > 10^0,$$

where A_1 is the combined atmospheric absorption, l is the total distance through the cloud, A_0 is the zenith attenuation of the clear atmosphere (from Figure 2.7),

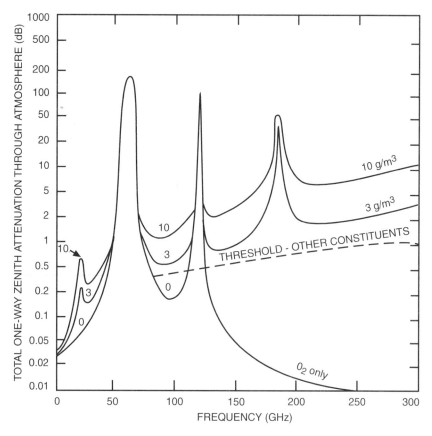

Figure 2.7   Total one-way zenith attenuation through the atmosphere as a function of water vapor density, 1 to 300 GHz.

and θ is the elevation angle. The combined brightness temperature is then obtained from expression (2.20) or Figure 2.6.

Rain may be treated in much the same way. The absorption due to rain as a function of rainfall rate has been calculated by several authors. Shown in Figure 2.8 is one such set [36]. It is then necessary to estimate an "effective path length" through the rain. Examples of such estimates are given in NASA's *Propagation Handbook* [37]. The estimation of rain attenuation has received a lot of attention in its own right and the interested worker is referred to this

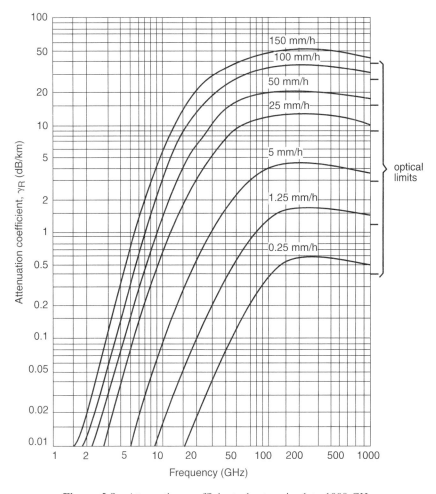

Figure 2.8   Attenuation coefficients due to rain, 1 to 1000 GHz.

literature [37,38]. Finally, the attenuation due to rainfall is added to that due to the gaseous atmosphere and to clouds as

$$A_2 = A_1 + \gamma_R d_e, \qquad (2.21)$$

where A_2 is total atmospheric absorption, γ_R is the attenuation due to rainfall (dB/km), and d_e is the effective distance through the rainfall. The resultant brightness temperature may now be obtained from Eq. (2.22) or Figure 2.6.

One point has been glossed over. Kirchoff's law relates emission to absorption and not to total attenuation. Particularly for rain, the attenuation will have a significant scattering component as frequency and rainfall rate increase, which complicates the problem. In most instances this will be a second-order effect [39,40].

The quantum limit to system performance is a consequence of the particle (photon) behavior of electromagnetic radiation. The effect of the ground on antenna sidelobes is quite complex. Common practice [22] is to assume a ground emission of, say, 290 K and to make some adjustment accordingly. In fact, the brightness temperature of the ground is

$$T_b = \varepsilon T_{surf} + \rho_r T_b' + \rho_s T_{sky}, \qquad (2.22)$$

where $\varepsilon = 1 - \rho$ is the emissivity of the ground, a function of frequency, polarization, ground constants, elevation of angle and roughness; $\rho = \rho_r + \rho_s$ is the power reflection coefficient; ρ_r is the power Fresnel coefficient; ρ_s is the power scattering coefficient; T_{surf} is the temperature of the ground; T_b' is the brightness temperature at the angle of the reflected ray; and T_{sky} is the sky noise temperature in the direction of scattered energy.

The resultant antenna temperature T_A may then be obtained from [20]

$$T_A = \frac{\int\limits_{\Omega} T_b(\Omega)\, g(\Omega)\, d\Omega}{\int\limits_{\Omega} g(\Omega)\, d\Omega}, \qquad (2.23)$$

where $d\Omega$ is the differential of the solid angle and $g(\Omega)$ is the antenna radiation pattern gain distribution.

2.9   Quiet and Disturbed Plasmasphere

The ionosphere of the earth extends deeply into the magnetosphere. In the F-region and above (i.e., in the exosphere), the distribution of the ionization is controlled by the magnetic field line distribution. The distribution of the electron and ion densities is organized along magnetic field lines. Ionospheric plasma density irregularities are stretched out along magnetic flux tubes to higher altitudes into the protonosphere–magnetosphere. The filaments of plasma form preferential ducts for very low frequency (VLF) whistlers wave propagation.

The cold ionospheric plasma (0.25–0.5 eV; 3000–6000 K) is gravitationally bound to the earth. Its density decreases above the F-region with a scale height which is proportional to the plasma temperature $(T_e + T_i)/2$, ("e" for electron, "i" for ions) and inversely proportional to the gravitational force (mg).

Upward diffusion and evaporation of charged particles contribute to replenish magnetic flux tubes as soon as they have been emptied at the onset of major geomagnetic perturbations. The maximum upward refilling flux is found to be of the order of 3×10^8 ions and electrons per cm^2 and per second. Empty flux tubes at low latitude and midlatitude are refilled in less than a week with new plasma pouring out of the terrestrial atmosphere. This means that the ionization density in plasmaspheric flux tubes at $L < 5$ reaches saturation level— corresponding to diffusive and hydrostatic equilibrium—in less than 6–7 days.

Catastrophic depletions of the plasmasphere are observed during large geo-magnetic perturbations which often occur before the saturation level has been reached. The equatorial plasma density then drops from a near-saturation value (300–500 cm^{-3}) to less than 10 cm^{-3}, in a rather short period of time starting at the onset of the geomagnetic substorm. The portion of the plasmasphere which is then peeled off depends on the strength of the geomagnetic perturbation as measured, for instance, by the Kp index. During large storms, the plasmasphere can be depleted and peeled off along geomagnetic field lines as low as $L = 2$. However, such deep depletions are relatively rare events. Therefore, flux tubes at $L < 3$–4 are most of the time close to saturation level, while those beyond $L = 4$ are usually in a dynamical state of refilling. When the level of geomagnetic activity is steady for 24 h or more, the thermal ion and electron densities remain almost unperturbed in all geomagnetic flux tubes located inside $L = 4$. When magnetic agitation has been at the same level for a day or more, and increases subsequently, a well-developed sharp boundary is formed in the plasmasphere. This surface was called "plasmapause" by D. L. Carpenter [41], who discovered it from VLF whistler observations. The sharp density gradient forcing at the plasmapause surface has since been observed with many different satellites [42–45]. Actually, the plasmapause density "knee" had first been observed in 1960 by Gringauz with ion traps [46,47].

At a recently formed plasmapause, the equatorial density decreases very abruptly by two orders of magnitude from 300–500 cm^{-3} to less than 10 cm^{-3}, over an equatorial distance of $0.15\,R_{\mathrm{E}}$ (R_{E} is earth radius). These density gradients separate partly empty magnetospheric flux tubes just outside the plasmapause and those which are in the process of refilling just inside this boundary.

A three-dimensional representation of the plasmasphere and of its outer boundary is illustrated in Figure 2.9. The equatorial cross-section of the doughnut-shaped surface is a function of local time (LT). For steady and moderate geomagnetic activity the plasmapause has a bulge extending to $L = 6\text{–}7$ in the dusk region.

During prolonged very quiet geomagnetic conditions, the plasmasphere has a tendency to fill maximum space in the magnetosphere. The sharp equatorial density "knee" formed during the latest magnetospheric substorm onset has the irreversible tendency to smooth out and to disappear gradually during the following prolonged quiet period of time. The plasmasphere relaxes then to a more axisymmetric shape, with, however, a characteristic bulge in the noon local time sector. However, when geomagnetic activity increases, the nearly symmetrical plasmasphere is compressed in the post-midnight local time sector, while, on the dayside, the thermal plasma is expanded in the sunward direction. A new plasmapause gradient is then formed in the post-midnight local time sector at an equatorial distance which is approximately given by

$$L_{\mathrm{pp}} = 5.7 - 0.43\,(K_{\mathrm{p}})^{-12},$$

where $(K_{\mathrm{p}})^{-12}$ is the maximum value of the geomagnetic index K_{p} during the 12 preceding hours [48]. Once formed in the nightside region, the new density "knee" corotates toward dawn and toward later LT hours, as illustrated in Figure 2.10. In the dayside local time sector, the equatorial position of the plasmapause is determined by the level of activity at an earlier Universal Time, that is, when the corresponding plasma element was convecting past the post-midnight LT sector (see Figure 2.10) [45,49,50]. While the new density gradient formed near midnight propagates toward later local-time hours, its sharpness gradually decreases to become spread over much broader radial distances in the afternoon LT sector.

Following short-duration K_{p} enhancements, small plasmasphere bulges are formed in noon local sector as a result of the sunward plasma drift associated with enhanced dawn–dusk component of the magnetospheric electric field. Subsequently, these dayside bulges corotate toward the dusk local time sector as K_{p} decreases. Detached plasma elements or plasma tails are often observed in the afternoon local time sector [51], as well as in the post-midnight sector.

At altitudes below 3000 km, the signature of the plasmapause is not always clearly identifiable. There are, however, in the topside ionosphere, different

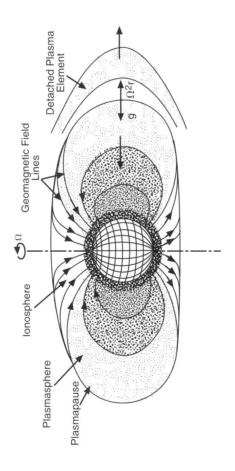

Figure 2.9   Three-dimensional illustration of the plasmasphere and its outer boundary: the plasmapause.

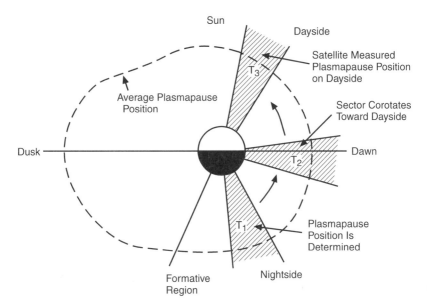

Figure 2.10   An illustration of plasmapause formation in the nightside region.

features which are related to the equatorial plasmapause. These lower altitude features are (1) the midlatitude electron density trough which was discovered by Muldrew [52], (2) the light ion trough (LIT), first named by Taylor *et al.* [53], (3) the plasmapause-associated temperature enhancement, and also (4) Stable Auroral Red arcs (SAR) [54], which are observed in the vicinity of the footprints of plasmapause field lines. A series of satellite observations have confirmed that the light ion trough in the topside ionosphere is consistently located at about one *L*-value smaller than the equatorial plasmapause magnetic field lines. Furthermore, the LIT does not exhibit well-developed dawn–dusk or noon–midnight asymmetries like the equatorial plasmapause. It must also be mentioned that significant polar-wind-like upward ionization flow is not only observed outside the plasmapause surface, but also in the intermediate region inside the plasmasphere [42,55]. In this outermost portion of the plasmasphere the electron temperature is usually much higher than the corresponding ionospheric temperatures.

The upward ionization flow is predominantly composed of suprathermal H^+ ions with 10% He^+ ions of an energy ranging between 1 and 2 eV [56]. Finally, it is worthwhile to point out that beyond the outer edge of the plasmapause the upward ion flux contains generally more suprathermal O^+ ions. From the large amount of observations collected since 1963, when this new magnetospheric frontier was discovered, it has become evident that the plasmasphere is a highly structured and variable body of corotating cold plasma. Steady-state models for the magnetospheric electric field and for the plasmasphere have often been used to describe the formation of this boundary. These stationary models have, however, only a limited usefulness in describing the proper physical processes involved in the formation of the equatorial plasma density knee. Time-dependent electric field models eventually had to be introduced to simulate in a more realistic way the dynamical motion and deformations of the plasmasphere as well as of its outer edge during periods of variable geomagnetic conditions.

Finally, it should be emphasized that the whole plasmasphere is magnetically and electrically coupled (connected) to the low- and midlatitude ionosphere. As a result of the finite value of the transverse Pedersen conductivity in the ionosphere, plasma interchange motion driven by various forces can develop only at finite velocity. The electric resistivity in the lower ionosphere limits the plasma interchange velocity to a maximum value. It is with this maximum velocity that blocks of plasma are detached from the plasmasphere in the post-midnight local time sector at the onset of each new substorm when the azimuthal convection velocity is suddenly enhanced. As a consequence of this detachment, a new plasma density gradient is formed in the inner magnetosphere along the zero radial force (ZRF) surface where the gravitational force is balanced by the enhanced centrifugal force [57,58]. A multiple-step equatorial density profile (see Figure 2.11) can be formed by a series of such plasmasphere peeling-off processes occurring in sequence, at larger and larger radial distances.

2.10 Tracking and Watching for Space Storms

In February 1997, the weather satellites watching the U.S. weather were themselves enduring space storms. A solar eruption apparently disturbed the attitude control of a Geostationary Operational Environmental Satellite (GOES-8) on Feb. 14. Electric discharges in the satellite's electronics induced four or five "bit flips" in the RAM that lets operators fine-tune the attitude control. Upsets and damage may have occurred in other satellites, as well—those used by military and commercial operators, who rarely disclose any problems.

Solar activity embraces a variety of conditions, including localized changes in the sun's temperature and in the strength or direction of its magnetic fields, and the eruption of gases and plasmas from its surface. All this creates a wind of electrons, helium and hydrogen atoms and ions, and other charged and uncharged

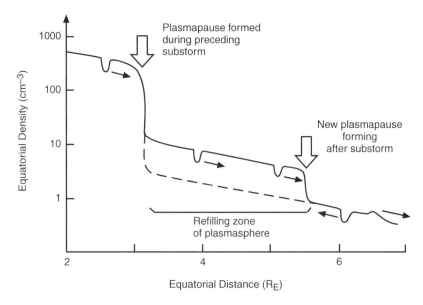

Figure 2.11   Illustration of the formation of multiple plasmapauses after an extended period of high geomagnetic activity.

particles, some of which have percolated from within the sun. This wind is thrown from the outer atmosphere of the sun and moves across the solar system (see Figure 2.12). The storms that affected the satellites were driven by the coupling of the earth's own magnetic field with this solar wind. Normally the wind streams at 400–700 km/sec outward along the plane of the solar system. Its intensity and output can vary with the 11-year cycle in which sunspots—dark areas on the sun's surface—wax and wane.

Effects from those space storms can swamp satellite electronics and disrupt electrical power and communications on earth. Nevertheless, the space weather stations that monitor such solar activity are surprisingly few. The National Oceanic and Atmospheric Administration (NOAA) operates two sets of weather satellites, the Geostationary Operational Environmental Satellites (GOES) in geostationary orbit, 36,000 km above the surface of the earth, and NOAA satellites in pole-crossing orbits, 840 km high. Two major programs now under way promise to help solve the problem of predicting potentially damaging solar activity. Under the International Solar–Terrestrial Physics (ISTP) program, the United States, Europe, Japan, and Russia are providing satellites that will plug gaps in people's

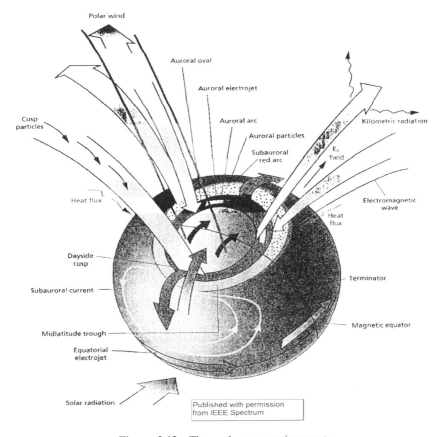

Polar wind

Auroral oval

Auroral electrojet

Cusp
particles

Auroral arc

Kilometric radiation

Auroral particles

Subauroral
red arc

E∥
field

Heat flux

Electromagnetic
wave

Heat
flux

Dayside
cusp

Terminator

Subauroral current

Midlatitude trough

Magnetic equator

Equatorial
electrojet

Solar radiation

Published with permission
from IEEE Spectrum

Figure 2.12   The earth space environment.

knowledge of what happens as the solar wind sweeps past the earth and pumps
energy into the magnetosphere. This region is a comet-shaped bag of plasmas
(ionized gases) and energetic particles that are held captive by the earth's magnetic
field (see Figure 2.13). At this writing, two spacecraft, Geotail and Wind, were
in orbit, and two more were soon to join them. Another program is the National
Space Weather Program initiative, which is run by a committee under the Federal
Coordinator for Meteorology, Washington, D.C. The group is forming an inter-
agency network to coordinate observations and research using these satellites
and to develop reliable forecasts of potentially damaging solar and geomagnetic
activity.

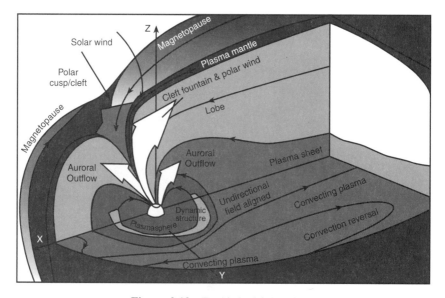

Figure 2.13   Earth's invisible shield.

2.11   Space Weather Channel

For now, warnings and forecasts of solar activity and its effect on the magneto-sphere are provided to civilian and industrial customers by the Space Environmental Laboratory (SEL)'s Space Environment Services Center, which is staffed on two shifts by forecasters and around the clock by solar technicians who watch their instruments for changes on the sun. These changes include sunspots (darker regions on the sun's surface caused by intense magnetic activity), solar flares (visible, energetic eruptions of gas extending thousands of kilometers from the solar surface), and changes in the sun's hot outer atmosphere known as the corona. Every 24 hours, SEL produces one summary of solar activity and another of geophysical activity. Solar activity covers occurrences on the surface of the sun, whereas geophysical activity covers changes in conditions in the space and atmosphere around the earth. Each report describes what actually happened over the past 24 hours and tries to predict what will happen in the next 72 hours.

As the world center for civilian warnings of space environmental activity, the SEL's space environment center collates data and reports from Moscow, Medoun (France), Tokyo, Sydney, and other locations. The Space Environment Services Center also gathers information from the Air Force's Solar Electro-Optical Network, which has five stations around the world that watch for indications of flares in the hydrogen-alpha spectral line (653.6-nm wavelength) emitted by hydrogen gas in the outer solar atmosphere. It is on this line, near the red end of the visible spectrum, that most of the activities on the visble surface of the sun can be seen. SEL forecasters also make use of magnetographs that apply the Zeeman effect to measure the intensity of magnetic fields in the solar atmosphere (magnetic fields excite atoms to emit light in multiple spectral lines where one line normally would appear). With these they can track regions where energy bound within magnetic structures is likely to cause eruptions, such as solar flares. Sunspots, the key feature associated with changes in the sun's magnetic field, are tracked with white-light telescopes. Other equipment at the center includes three radio telescopes in charge of sweep-frequency interferometers that measure electromagnetic radiation in seven radio bands ranging from 245 to 15,000 MHz for early signs of solar eruptions. Another interferometer, operating at 25–75 MHz, can see some materials leaving the surface of the sun toward the earth. A worldwide network of 30 magnetometers sends back data, nearly in real time, of changes in the magnetic field at the earth's surface. SEL also has a hydrogen-alpha telescope and solar X-ray monitors aboard GOES satellites.

When certain physical conditions such as solar X-ray and ultraviolet intensities exceed preset thresholds, SEL sends alerts to specific customers, starting with satellite and power grid operators. SEL also briefs NASA three times a day during Space Shuttle missions because solar flares could blast astronauts, even in their spacecraft, with dangerous doses of protons and other radiation. Data is also passed on to the Federal Aviation Administration to help airline crews flying at high altitude across the Arctic Circle gauge radiation. A complementary operation is provided by the Air Force's 50th Weather Squadron, Colorado Springs, Colorado. The squadron tracks data out of concern with how the ionosphere affects both the propagation of radar and Loran navigation signals and the arrival times of signals from Global Positioning System satellites.

When the ionosphere is heated by increased solar ultraviolet light or because of intense bombardment of the magnetosphere by radiation—called a geomagnetic storm—it expands and its electron content changes. These changes can affect how the ionosphere reflects radio signals from the ground or refracts signals from satellites, and thus skew the difference in radio signal arrival times that is vital to determining position.

A severe geomagnetic storm can also induce high voltage differences in any long conductor, such as power grids with electric lines stretching partway across the country. The reactions of electric utility companies to alerts depend on their location. They range from simply observing events to curtailing maintenance schedules or bringing up additional generating plants so that more generators are available for greater voltage regulation.

Even though forecasts are not completely reliable, "nowcasts"—brief alerts to current activity in the magnetosphere—help. Just knowing that a geomagnetic disturbance was to blame for a fault can save the cost of inspecting a transformer, a job that could take several days. On the other hand, storms can induce currents in a transfomer's primary windings that saturate and heat the core, degrading insulation and setting the stage for a later breakdown that may seem unrelated to any specific event. Large transformers cost approximately $10 million to replace, and buying electric power until a replacement unit is installed can cost $400,000 a day.

2.12   Effects on Satellites

The increasing sophistication and density of the electronics placed aboard satellites makes them more vulnerable to space storms. Satellites in the 1960s were less vulnerable because their electronics were larger and more robust. Now, with device dimensions of a few micrometers, solid-state electronics approach the same physical dimensions as the region of damage that can be caused by ionizing radiation, and they carry charges with magnitudes close to that generated by radiation particles striking silicon. During times of heightened solar activity, three principal elements of the space environment attack satellites with increased vigor. Ambient plasmas charge spacecraft surfaces and cause arc discharges across the vehicle. High-energy electrons penetrate deep into a spacecraft to build high charges in insulation, such as is present on coaxial lines. And protons and other charged particles disrupt computer memories or even damage the structure of semiconductor microelectronics. Ironically, the geostationary orbit, which is so valuable for communications and weather satellites, lies within the Van Allen radiation belt, which comprises three doughnut-shaped rings of charged particles and plasmas trapped by the earth's magnetic field. Because high-energy protons and electrons destroy solar cells, satellites are always launched with far more solar cells than they need at first, so enough cells will be left undamaged to power the craft as it nears the end of its mission. Solar protons, with energies ranging from 20 keV to 1 MeV or more, have easy access to the geostationary region. In just one event, they can do as much damage to solar cells as several years' exposure to energetic electrons. More seriously, high-energy electrons can penetrate deep into a spacecraft and build large static charges up to 19,000 V.

Shielding spacecraft against radiation is a nontrivial task, since it adds weight, often at the expense of the propellant needed to maintain the craft's orbit or attitude. But less propellant means a shorter useful satellite life, prematurity that can occur in unexpected ways. For example, on Jan. 20–21, 1994, a geomagnetic storm destroyed part of the electronics controlling the reaction wheels that stabilize Canada's Anik E-I communications satellite. Therefore, that function is now served by thruster rockets that were originally meant for infrequent orbital adjustments. The net result is that Anik E-1 will run out of fuel and need replacing 6 years earlier than planned.

One of the instruments on the U.S. Polar spacecraft will carry excess memory, so that enough will be left to make measurements with high time resolution even after several months of radiation damage.

Even satellites at lower altitudes are hurt by solar radiation. Ultraviolet light heats and expands the earth's upper atmosphere, which increases the drag on satellites. This change must be factored into reentry calculations for the Space Shuttle; also, in 1978–79, it accelerated the demise of the U.S. Skylab space station. The principal driver in magnetospheric physics is the sun, which, despite thousands of years of observations, still holds many secrets. The huge cauldron of superhot gases has its own magnetic field, which changes over time and space and which is largely responsible for the sun's observed activities.

For decades, the search for the origins of space storms has focused on solar flares. But recently, ejections of large quantities of plasma, called coronal mass ejections (CMEs), have come under scrutiny as a possible cause of the storms. The corona is a superhot shroud of gas around the outer atmosphere of the sun, where temperatures reach a million kelvins. It is usually enslaved to the sun's magnetic fields, but in 1973–74, X-ray telescopes carried aboard the U.S. Skylab space station revealed a new phenomenon: coronal ''holes.'' Skylab's X-ray images showed darkened areas through which the solar wind could blow outward, unimpaired. More recently, since 1991, X-ray telescopes aboard Japan's Yolikoh satellite have also revealed CMEs, which carry their own magnetic fields and can measure thousands of kilometers across.

Chapter 3 | Electromagnetic Interference and Receiver Modeling

3.0 Non-average Power Sensitivity Receptor Modeling

A detailed discussion of the waveform parameters total energy, peak current (and voltage), and rise time is given in this chapter with the help of research done for Refs. [59–67]. The discussion is in terms of the electromagnetic interference (EMI) margin for each of these parameters that preserves the important features of the average power margin. In particular all margins are in terms of readily measurable quantities, such as power spectral density calculated at the receptor's input; applicable to both stochastic and deterministic waveforms; and adhere to a "worst case" philosophy. These margins are in terms of quantities that utilize both input data and additional input data that is realistic and easily obtainable on a given system.

In the following sections, examples of EMI margins are developed. The discussions consider total energy, peak waveforms, and rise-time margins. These are presented as candidates and are discussed in terms of the general receptor model shown in Figure 3.1. Although this model considers only current waveforms the extension to voltage waveforms is straightforward and algorithms applicable to voltage waveforms can be developed. Table 3.1 contains a listing of parameters and their corresponding definitions used in this section.

3.1 Total Energy—Deterministic Waveforms

The total energy of a periodic waveform is infinite. Thus, such a waveform will always cause interference to an energy-sensitive receptor. However, in practice

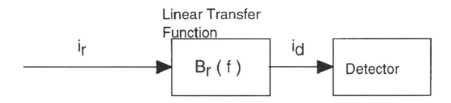

Figure 3.1 General receptor–receiver model.

Table 3.1 **Definition of Variables**

P_d	=	average power at input to detector (watts)
P_r	=	average power at input to receptor (watts)
E_d	=	total energy at input to detector (watts-sec)
$i_d(t)$	=	detector input current (amps)
$i_r(t)$	=	receptor input current (amps)
$i_r(t)_{peak}$	=	peak value of $i_r(t)$ (amps)
$I_d(t)$	=	Fourier transform of $i_d(t)$ (finite energy) (amps/Hz)
$I_r(f)$	=	Fourier transform of $i_r(t)$ (finite energy) (amps/Hz)
$I_r^s(f)$	=	level of $I(f)$ which induces the interference threshold (amps) level at the detector
K^P	=	detector interference threshold power level (watts)
KE	=	detector interference threshold energy level (watts-sec)
K	=	detector interference threshold peak current level (amps)
K^0	=	detector interference threshold bandwidth (Hz)
$G_r(f)$	=	spectral power density at receptor input (watts/Hz)
$B_r(f)$	=	receptor input-to-detector linear current/voltage transfer function
$|B_r(f)|^2$	=	receptor input-to-detector energy transfer function
$\Delta_r(f)$	=	time interval assigned to an energy sensitive (sec)
Δ	=	duration of interference on receptor (sec)
σ_d^2	=	variance of detector input waveform (watts)
σ_r^2	=	variance of receptor input waveform $2\int_{f_a}^{f_b} G_r(f)\,df$
α	=	fraction of time that a stochastic waveform peak at detector input must exceed K to trigger interference
f_p	=	frequency for which $B_r(f)$ is maximum (Hz)
f_o	=	center frequency of a narrowband Gaussian process (Hz)
f_a, f_b	=	lower, upper frequency defining common frequency band between interference and receptor (Hz)
s	=	amplitude of the sinusoid in a narrowband (volts or amps) Gaussian plus sinusoid process
τ	=	pulse width of a pulse interfering waveform (sec)
β	=	3-dB point bandwidth of $B_r(f)$ (Hz)
β_r^s	=	receptor input waveform bandwidth which induces the interference threshold bandwidth at the detector (Hz)
β_r	=	portion of the receptor input waveform bandwidth within the passband of $B_r(f)$ (Hz)
$\Omega(x)$	=	modified Bessel function of zero order

this interference cannot occur unless the average power exceeds the average rate of energy dissipation (e.g., heat loss due to environmental cooling). Thus, for periodic waveforms a total energy EMI criterion should actually be an appropriate power EMI criterion. The present average power EMI margins are directly applicable for this case.

For a periodic waveform, the total energy is defined as

$$E = \int_{t_1}^{t_2} f^2(t) \, dt, \tag{3.1}$$

where (t_1, t_2) is the time interval of the waveform and it is understood that a reference of 1 ohm is used. If $f(t)$ satisfies the condition

$$\int_{\infty}^{\infty} f^2(t) \, dt < \infty, \tag{3.2}$$

it is said to have finite energy and is called an energy signal.

Recognizing that a nonperiodic function may be represented by the Fourier transform pair, we have

$$f(t) = \int_{\infty}^{\infty} F(f) e^{j\omega t} \, df \tag{3.3}$$

and

$$F(f) = \int_{\infty}^{\infty} f(t) e^{-j\omega t} \, dt. \tag{3.4}$$

Using the transform pair and the above relationship, we can form

$$\int_{\infty}^{\infty} f^2(t) \, dt = \int_{\infty}^{\infty} f(t) \left[\int_{\infty}^{\infty} F(f) e^{j\omega t} \, df \right] dt, \tag{3.5}$$

and by inversion of the order of integration we have

$$\int_{\infty}^{\infty} f^2(t) \, dt = \int_{\infty}^{\infty} |F(f)|^2 \, df. \tag{3.6}$$

This result states that the total energy in a given nonperiodic time function is simply the area of the $|F(f)|^2$ curve. The term $|F(f)|^2$ is called the energy-density function and expresses the energy of $f(t)$ as a function of frequency. Thus, $|F(f)|^2$ has the units of watt-sec/Hz.

As an example, consider a signal having an arbitrary energy spectrum passed through an ideal bandpass filter centered at frequency f_1. Assume that the energy transfer function of the filter is unity for components lying in the filter passband and zero for other components (Figure 3.2). The total energy of the output is

$$E_o = \int_0^\infty |V_o(f)|^2 \, df = \int_{f_1 - W/2}^{f_1 + W/2} |V_i(f)|^2 \, df. \tag{3.7}$$

For a sufficiently narrow filter bandpass (narrow enough so that the input spectrum is essentially constant over the band), the output can be approximated as

$$|V_i(f_1)|^2 = \frac{E_o}{W}. \tag{3.8}$$

From this expression it is evident that $|V_i(f_1)|^2$ can be interpreted as the energy per unit bandwidth.

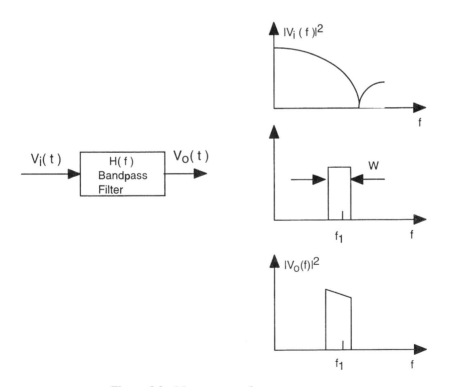

Figure 3.2 Measurement of energy spectrum.

To determine the energy spectrum of a periodic function, consider the pulse shown in Figure 3.3a. This pulse can be expressed analytically as

$$P_T = 1, \qquad \frac{-T}{2} < t < \frac{T}{2}$$
$$= 0 \qquad \text{otherwise.}$$

The Fourier transform is obtained from

$$P_T(\omega) = \int_{\infty}^{\infty} P_T(t)e^{j\omega t}\,dt = \int_{T/2}^{T/2} e^{j\omega t}\,dt = \frac{e^{j\omega t}}{-j\omega}\Bigg|_{-T/2}^{T/2} = \frac{e^{j\omega T/2} - e^{-j\omega T/2}}{j\omega}. \qquad (3.9)$$

Converting the exponentials to the equivalent trigonometric functions leads to

$$P_T(\omega) = T\frac{\sin(\omega T/2)}{\omega T/2}.$$

The energy spectrum of the pulse signal, $P_T(t)$, is

$$|P_T(f)|^2 = 2T^2\left(\frac{\sin \pi T f}{\pi T f}\right)^2 = 2T^2\,\text{sinc}^2(fT). \qquad (3.10)$$

The energy spectrum $|P_T(f)|^2$ of the rectangular pulse is shown in Figure 3.3b. It is seen that the energy is concentrated in the low-frequency portion of the spectrum. The extent of this concentration can be found by computing the energy in the first loop (that is, for $|f| < 1/T$) and comparing this to the total energy. The ratio, found by graphical integration, is 0.902. Thus, 90.2% of the energy in a rectangular pulse is contained in the band of frequencies below a frequency

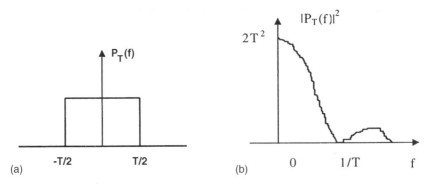

(a)   (b)

Figure 3.3   Energy spectrum of a periodic function: (a) pulse; (b) spectrum.

equal to the reciprocal of the pulse length. As a rule of thumb, it is often assumed that a pulse transmission system having a bandwidth equal to the reciprocal of the pulse width will perform satisfactorily. Actually, if high-fidelity reproduction of the pulse shape is required, a much greater bandwidth will be necessary. However, it can be seen that a system with this bandwidth will transmit most of the pulse energy. Using Eq. (3.1), the total energy at the input to the detector (on a 1-ohm basis) is given by

$$E_d = \int_0^\infty |I_d(f)|^2 \, df, \qquad \text{(watt-sec)}$$

where $I_d(f)$ is the Fourier transform (one-sided) of the detector input waveform $i_d(t)$.

For a simple system with system function $B_r(f)$, the output and input are related by

$$I_d(f) = B_r(f)I_r(f),$$

where $I_r(f)$ is the Fourier transform of the receptor input waveform $i_r(t)$. Thus, the energy spectrum of the output is

$$|I_d(f)|^2 = I_d(f)I_d^*(f) = |B_r(f)|^2 \, |I_r(f)|^2 \qquad (3.11)$$

and

$$E_d = \int_0^\infty |B_r(f)|^2 \, |I_r(f)|^2 \, df. \qquad (3.12)$$

The energy susceptibility function may be represented as shown in Figure 3.4. As in the case of the power susceptibility curve, the energy susceptibility curve

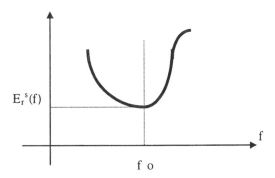

Figure 3.4   Energy susceptibility function.

is a minimum at that frequency where the transfer function $B_r(f)$ is a maximum. The detector interference threshold energy level K^E can be related to the receptor input energy for a sinusoidal waveform in a time interval, $\Delta_r(f)$. Thus, the susceptibility energy is

$$K^E = B_r(f)^2 E_r^s(f), \tag{3.13}$$

where $E_r^s(f)$ is the CW energy at frequency f which generates the energy equal to the standard response energy level at the detector input. The total energy interference margin for a deterministic, finite-energy waveform becomes

$$\frac{E_d}{K^E} = \int_{f_b}^{f_a} \frac{|I_r(f)|^2}{E_r^s(f)}\, df, \tag{3.14}$$

where (f_a, f_b) are the frequency limits for the energy-susceptible devices. Note that the measurable quantities are transferred to the receptor input.

The energy received at the receptor from an emitter is given by the area under the received energy density function times the input impedance of the receptor. The result is

$$|I_t(f)|^2\, t(f)\, br_{ir}, \tag{3.15}$$

where $|I_t(f)|^2$ is the transmitted energy density (watt-sec/Hz, 1 ohm), b is the bandwidth factor for the emitter (Hz), $t(f)$ is the transmission loss, and r_{ir} is the input impedance of the ith receptor (ohms). This assumes $|I_t(f)|^2$ is a constant over bandwidth b. This last term can also be expressed as $|I_t(f)|^2 = (q\sqrt{2})^2$, where q is the current spectral level (amps/Hz). The broadband point energy interference margin is then given by

$$\text{epm}_d(f) = \frac{(\sqrt{2}q)^2\, t(f)br_{ir}}{E_r^s(f)}. \tag{3.16}$$

Converting to decibels, the broadband point energy interference margin for an aperiodic signal becomes

$$\begin{aligned} \text{EPM}_D(f)_{dB} = {} & Q\ (\text{dB}\ \mu A / \text{MHz}) + T(f)\ (\text{dB}) + B(\text{dB MHz}) \\ & + R_{IR} - E_r^s(f) - I_r^s(f)(\text{dB}\ \mu A), \end{aligned}$$

where $\text{EPM}_D(f)(\text{dB}) = 10\ \log\ \text{epm}_d(f)$, $Q(\text{dB}\ \mu A/\text{MHz}) = 20\ \log (q\sqrt{2}/10^{-12})$, $T(f)(\text{dB}) = 10\ \log\ t(f)$, $B(\text{dB MHz}) = 10\ \log\ (b/10^6)$, $R_{IR}(\text{dB}) = 10\ \log\ r_{ir}$, and $E_r^s(f)(\text{dB}) = 10\ \log\ E_r^s(f)$.

3.2   Total Energy—Stochastic Waveforms

There may be instances where certain emitters are considered sources of "switched" stochastic waveforms in that an otherwise stationary process is turned on at known intervals. For example, consider a rotating reflector antenna that is emitting narrowband Gaussian noise within a receptor bandwith. The total energy at the detector of the receptor can be determined from

$$E_d = \Delta P_d,$$

where Δ is the duration of interference on the receptor and where P_d is given by

$$P_d = \int_0^\infty G_r(f)|B_r(f)|^2 \, df. \tag{3.17}$$

As in the deterministic case, the detector interference threshold energy level is

$$K^E = |B_r(f)|^2 \, E_r^s(f).$$

It follows that the total energy EMI margin for the "switched" stochastic waveform is

$$\frac{E_d}{K^E} = \Delta \int_{f_a}^{f_b} \frac{G_f(f)}{E_r^s(f)} \, df, \tag{3.18}$$

where (f_a, f_b) is the frequency limit for the energy-susceptible device.

A value of broadband emitter power spectral density is assigned to each sample frequency. The broadband power received at a receptor from an emitter is given by the area under the received power spectral density times the input impedance of the receptor. This may be defined as

$$G_r(f)* \, (\text{bw})* \, r_{ir}$$

where

$G_r(f) = G_t(f) \cdot t(f)$
$G_t(f) = $ emitter power density at frequency f (watts/Hz)
$(\text{bw}) = $ bandwidth (Hz) and $t(f)$ and r_{ir} are as defined previously.

The broadband energy point of interference margin (switched stochastic) is then given by

$$\text{epm}_s(f) = \frac{(\Delta G_r(f)(\text{bw})r_{ir}}{E_r^s(f)} = \frac{(\Delta)G_t(f)t(f)(\text{bw})}{E_r^s(f)}. \tag{3.19}$$

Converting to decibels, the broadband point energy EMI margin for the switched stationary signal becomes

$$\text{EMP}_s(f)(\text{dB}) = \Delta(\text{dB sec}) + Q(\text{dB } \mu\text{A}/\text{MHz}) + \text{BW}(\text{dB MHz}) \tag{3.20}$$
$$+ T(f)(\text{dB}) - \Delta_r(f)(\text{dB sec}) - I_r^s(f)(\text{dB } \mu\text{A}),$$

where

$\text{EMP}_s(f)(\text{dB})$	$= 10 \log \text{epm}_s(f)$
$\Delta(\text{dB sec})$	$= 10 \log \Delta$
$Q(\text{dB } \mu\text{A/MHz})$	$= 20 \log (q/10^{-12})$
$\text{BW}(\text{dB MHz})$	$= 20 \log (\text{bw}/10^6)$
$T(f)(\text{dB})$	$= 10 \log t(f)$
$\Delta_r(f)$	$= 10 \log \Delta_r(f)$
$I_r^s(f)(\text{dB } \mu\text{A})$	$= 20 \log (I_r^s(f)/10^{-6})$.

The integrated energy EMI margin for the switched stationary case is determined in the same manner as that presented in the deterministic total energy case.

3.3   Peak Current/Voltage—Deterministic Waveforms

Some receptors (e.g., many digital devices) are sensitive to the peak value of a waveform (such as voltage or current). An upper bound to this peak can be given in terms of amplitude spectral density frequency-domain data. This bound can be used to define a conservative estimate of a peak current (or voltage) EMI margin for deterministic waveforms in terms of receptor input quantities. The remainder of this chapter pertains to peak current, but the peak voltage deviation can be performed in an analogous manner. Consider the detector current given by

$$i_d(t) = \int_{\infty}^{\infty} I_r(f) B_r(f) e^{j\omega t} \, df. \tag{3.21}$$

Note that $I_r(f)$ is a superposition of impulses for periodic (infinite-duration) waveforms and a continuous function for finite-energy (finite-duration) waveforms. The detector interference threshold peak current level K is given in terms of a CW receptor input level $|I_r^s(f)|$ by

$$K = |B_r(f)| \, |I_r^s(f)|.$$

An interference margin may be defined by

$$\left| \frac{i_d(t)}{K} \right| \leq \int_{f_a}^{f_b} \frac{|I_r(f)|}{|I_r^s(f)|} \, df, \tag{3.22}$$

where (f_a, f_b) is the frequency limit for the peak current susceptible device. The peak current interference margin for deterministic waveforms is defined by

$$\int_{f_a}^{f_b} \frac{|I_r(f)|}{|I_r^s(f)|} \, df.$$

The peak current susceptibility function may be represented as shown in Figure 3.5. As in the case of the power susceptibility curve, the peak current susceptibility curve is a minimum at that frequency where the transfer function $B_r(f)$ is a maximum. The detector interference threshold peak current level K can be related to the receptor input peak current for a sinusoidal waveform by

$$K = |B_r(f)| I_r^s(f),$$

where $I_r^s(f)$ is the peak CW input signal at frequency f needed to produce the standard response peak current level at the detector input.

To get an understanding of how "worst case" the preceding equation is, consider a rectangular pulse train as shown in Figure 3.6. Using this example, we will demonstrate how it would make use of the equation. Assume $|I_r^s(f)| = 1$ and determine $|I_r(f)|$. From Figure 3.7, the power density is given by

$$
\begin{aligned}
P_{BB}(f) = 2A^2\tau^2 f_B &= 2 \times (.1)^2 \times (1.25 \times 10^{-2}) \times 4.0 \times 10^3 \\
&= 1.25 \ \mu\text{W/Hz} \qquad \text{for} \qquad 0 \le f \le f_m \\
= 1.25 \times 10^{-6} \ (f_m/f)^2, &\qquad f > f_m
\end{aligned}
\tag{3.23}
$$

$$\text{Bandwidth} = 1/2\tau = 1/(2 \times 1.25 \times 10^{-4}) = 4 \text{ kHz}$$
$$f_m = 1/\pi\tau = 1/(\pi \times 1.25 \times 10^{-4}) = 2.55 \text{ kHz}.$$

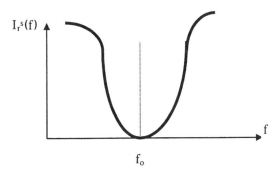

Figure 3.5   Peak current susceptibility function.

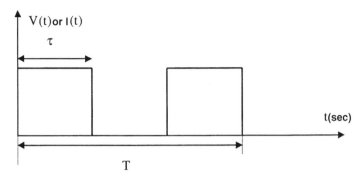

Figure 3.6   Rectangular pulse train.

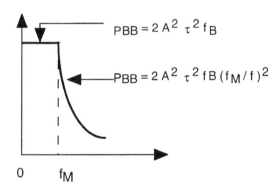

A= peak current/voltage into 1 ohm
$f_M = 1/\pi\tau$
bandwidth $= 1/2\tau$
$\tau =$ pulse width
$f_B =$ bit rate

Figure 3.7   Spectrum of rectangular pulse.

The Fourier series representation for the rectangular pulse train can be determined from

$$i(t) = \sum_{n=-\infty}^{\infty} \alpha_n\, e^{jn\omega_0 t},$$

where

$$\alpha_n = \frac{1}{T} \int_{t_1}^{t_1+T} i(t) e^{-jn\omega_0 t}\, dt,$$

where T is the period of pulse train. The coefficients are

$$\alpha_n = \frac{a\tau}{T}\left[\frac{\sin\dfrac{n\pi\tau}{T}}{\dfrac{n\pi\tau}{T}}\right]\exp\left[-j\frac{2\pi n}{T}\left(\frac{\tau}{\alpha}\right)\right], \tag{3.24}$$

where $\omega_0 = 2\pi/T$. Thus, the complete Fourier series expression or $i(t)$ now becomes

$$i(t) = \frac{a\tau}{T} + \frac{2a\tau}{T}\frac{\sin(\pi\tau/T)}{n\pi\tau/T}\cos\frac{2\pi n}{T}(t - \tau/a) \tag{3.25}$$

For the rectangular pulse (Figure 3.4)

$$P_{BB}(f) = 2A^2\tau^2 f_B, \qquad\qquad 0 < f < f_M \text{ (W/Hz)}$$
$$P_{BB}(f) = 2A^2\tau^2 f_B\,(f_M/f)^2, \qquad f > f_M,$$

where bandwidth $= 1/2\tau$, f_B is the bit rate, A is the peak current/voltage into 1 ohm, $f_M = 1/\pi\tau$, and τ is the pulse width.

The values in Figure 3.6 are:

$$A = 0.1$$
$$\tau = 1.25 \times 10^{-4} \text{ sec}$$
$$T = 0.25 \times 10^{-3} \text{ sec}$$

When a current $i(t)$ flows through a one ohm resistor, the power dissipated is

$$P = \langle i^2(t)\rangle,$$

where

$$\langle i^2(t)\rangle = \frac{1}{T}\int_{-T/2}^{T/2} i^2(t)\,dt$$

for a periodic signal. Hence, the power in a rectangular pulse train is given by

$$P\left(\frac{n}{T}\right) = \frac{2a^2\tau^2}{T^2}\left(\frac{\sin n\pi\tau/T}{n\pi\tau/T}\right)^2, \qquad \frac{n}{T} > 0. \tag{3.26}$$

To convert to a continuous spectrum, multiply $P(n/T)$ by T and we have

$$P_{BB}(n) = \frac{2a^2\tau^2}{T}\left(\frac{\sin n\pi\tau/T}{n\pi\tau/T}\right)^2, \tag{3.27}$$

where $P_{BB}(n)$ is the envelope of $P(n/T)$. Using the values for the preceding example,

$$P_{BB}(n) = \frac{2 \times (0.1)^2 \times (1.25 \times 10^{-4})^2}{0.25 \times 10^{-3}} \left(\frac{\sin n\pi \times \dfrac{1.25 \times 10^{-4}}{0.25 \times 10^{-3}}}{\dfrac{\pi \times 1.25 \times 10^{-4} \times n}{0.25 \times 10^{-3}}} \right)^2$$

$$= 1.25 \times 10^{-6} \left(\frac{\sin 90n}{\dfrac{\pi}{2} n} \right)^2.$$

A plot of $P_{BB}(n)$ is shown in Figure 3.8, which shows a plot of the interference model and the required frequency range. The required frequency range is a user-input option. To convert to a continuous current spectra, multiply $i(t)$ by T

$$i_e\left(\frac{n}{T}\right)\left(\frac{\text{amps}}{\text{H}_z}\right) = 2a\tau\left(\frac{\sin \pi n\tau/T}{\pi n\tau/T}\right), \qquad (3.28)$$

and with the preceding parameters:

$$i_e(n) = 2 \times 0.1 \times 1.25 \times 10^{-4}\left(\frac{\sin 90n}{\dfrac{\pi}{2} n} \right) = 25 \times 10^{-6}\left(\frac{\sin 90n}{\dfrac{\pi}{2} n} \right).$$

The representation for $i_e(n)$ is shown in Figure 3.9. From Figure 3.7, the model may be converted to current spectra by

$$I_r(f)\left(\frac{\text{amps}}{\text{Hz}}\right) = \sqrt{\frac{2}{f_B}P_{BB}(f)}, \qquad 0 \le f \le f_m$$

$$= \sqrt{\frac{2 \times 1.25 \times 10^{-6}}{4.0 \times 10^3}} \; (\text{amps/Hz}) = 25 \times 10^{-6}\left(\frac{f_m}{f}\right), \qquad f > f_m.$$

$$(3.29)$$

$I_r(f)$ is shown in Figure 3.9 for the above example. Figure 3.9 also shows the MIL-STD-461A port spectra. Using Figure 3.9, the effects of computing the peak current margin from Equation (3.2) may be determined. Recognizing that $i_e(n)$ is of the form $\sin x/x$, the integral (normalized to the peak value)

$$S_i(x) = \int_0^x \frac{\sin x}{x} \, dx$$

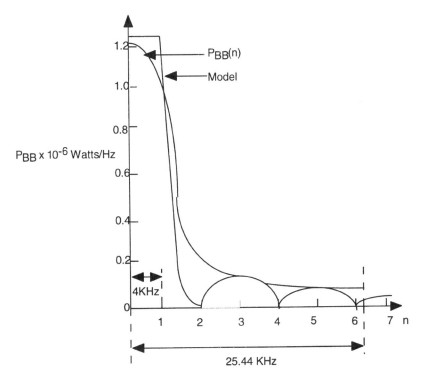

Figure 3.8 Plot of P_{BB} and interference model.

may be found in tabulated form in many different texts. Thus, the peak is defined by

$$i_{pe} = S_i(x)$$

where i_{pe} is the peak current associated with $i_c(x)$. The integral is given by

$$i_p = X_M \left(1 + \ln \frac{x}{x_M} \right),$$

where x_M is the value of x corresponding to f_m of the model.

The normalized models are shown in Figure 3.10 and the peak current calculation for Eq. (3.6) for various required frequency ranges are tabulated in Table 3.2. In Table 3.2, the column showing the ratio of the peak current (i_p/i_{pe}) represents the factor by which the interference model overpredicts the actual peak current for a rectangular pulse train may be overpredicted by as much as

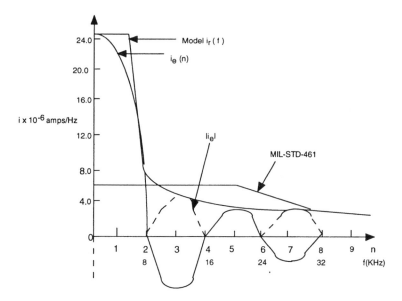

Figure 3.9   Plot of $i_e(n)$ model, $|i_e|$, and MIL-STD-461A.

a factor of 2. In terms of the rms currents (normalized to one microamp), the narrowband point interference margin is

$$M_p^N (f_\ell) = \frac{(I/10^{-6})^2 \, t_{ij}(f_\ell)}{(I_s/10^{-6})^2}, \qquad (3.30)$$

where $t_{ij}(f_\ell)$ is the power transfer function of coupling path between the jth emitter port and the ith receptor port. The term I_s is the receptor rms current equivalent to the power susceptibility level, and f_ℓ is the ℓth sample frequency. Converting to peak, we have

$$M_{PP}^N (f_\ell) = \frac{(\sqrt{2} \, I/10^{-6})^2 \, t_{ij} \, (f_\ell)}{(I_r^s/10^{-6})^2},$$

where I_r^s is the peak receptor current equivalent to the peak susceptibility level, and converting to dB we have

$$M_{PP}^N (f_\ell)(\text{dB}) = T_{ij}(f_\ell)(\text{dB}) + I(\text{dB } \mu\text{A}) - I_s(\text{dB } \mu\text{A}), \qquad (3.31)$$

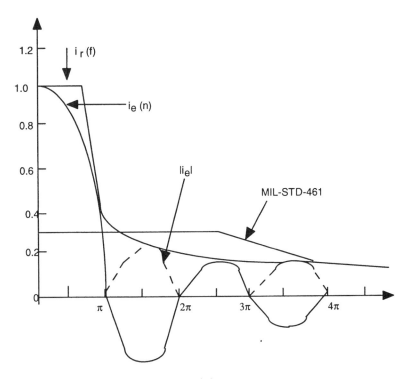

Figure 3.10   Normalized plot of $i_e(n)$, $|i_e|$, and MIL-STD-461 signal port spectra.

Table 3.2   **Comparison of Peak Current Calculations**

Required Frequency Range (kHz)	X	i_{pe} (amps/Hz)	$i_{p(model)}$ (amps/Hz)	$\dfrac{i_{p(model)}}{i_{pe}}$
4	1.57	1.36	1.45	1.07
8	3.14	1.85	2.14	1.17
12	4.71	1.61	2.55	1.58
16	6.28	1.42	2.84	2.0
20	7.85	1.56	3.06	1.96
24	9.42	1.67	3.24	1.94
28	11.0	1.58	3.40	2.15

where

$$M_{PP}^{N}(f_\ell) \quad = \quad 20 \log m_{PP}^{N}(f_\ell)$$

$$T_{ij}(f_\ell) \quad = \quad 10 \log t_{ij}(f_\ell)$$

$$I \text{ (dB } \mu\text{A)} \quad = \quad 20 \log \frac{\sqrt{2}\, I}{10^{-6}}$$

$$I_s \text{ (dB } \mu\text{A)} \quad = \quad 20 \log \left(\frac{I_r^s}{10^{-6}} \right).$$

The broadband emitter-current spectral density can be quantized by a computer program. A value of broadband emitter-current spectral density can be assigned to each frequency. This value is the maximum value assumed by the current spectral density within the corresponding frequency interval associated with the frequency. To evaluate the broadband current point interference margin at each sample frequency, we require the transfer function of the coupling path between an emitter and a receptor, and also a bandwidth factor, b. The bandwidth factor b is assigned to each sample frequency and is shown in Table 3.3.

The standard bandwidth (b_{std}) is associated with the EMC test instrument. The broadband peak current spectral level margin is determined by

$$\frac{|i_d(t)|_{max}}{k} \leq \int_{f_a}^{f_b} \frac{|I_r(f)|}{|I_r^s(f)|}\, df = \frac{|I_r(f)|b}{|I_r^s(f)|}, \tag{3.32}$$

where it is assumed that $|I_r(f)|$ and $I_r^s(f)$ are constants over b. The received peak current is

$$|I_r(f)| = |I_t(f)| \sqrt{t_{ij}(f)},$$

where $|I_t(f)|$ is the peak current spectral level at the emitter. Thus,

$$\frac{|I_r(f)|b}{|I_r^s(f)|} = \frac{|I_t(f)| \sqrt{t_{ij}(f)}\, b}{|I_r^s(f)|},$$

Table 3.3   **Bandwidth Factor**

Emitter	Receptor	Bandwidth
Required	Required	$\min(b_{emit}, b_{rec})$
Required	Nonrequired	$\min(b_{emit}, b_{std})$
Nonrequired	Required	b_{rec}
Nonrequired	Nonrequired	b_{std}

and the peak current broadband point margin for deterministic signals is defined by

$$m_{PP}^B(f) = \frac{|I_t(f)|\sqrt{t_{ij}(f)}\,b}{|I_r^s(f)|}.\tag{3.33}$$

The current spectral level is defined by

$$q = \frac{i_p/\sqrt{2}}{bw},$$

where i_p is the peak current. At the emitter

$$i_p = |I_t(f)|(bw),$$

and $q = |I_t(f)|/\sqrt{2}$.

It follows that $|I_t(f)| = \sqrt{2}q$. Therefore, the peak current broadband point margin may be expressed as

$$m_{PP}^B(f_\ell) = \frac{\left(\dfrac{\sqrt{2}q}{10^{-12}}\right)^2\left(\dfrac{f_B}{10^6}\right)t_{ij}(f_\ell)\left(\dfrac{b}{10^6}\right)}{\left(\dfrac{I_r^s}{10^{-6}}\right)^2} \times \frac{10^{-24}\times 10^6 \times 10^6}{(10^{-6})^2}.$$

Converting to decibels, the broadband point interference margin for a periodic signal becomes

$$M_{PP}(f_\ell)(dB) = Q(dB\ \mu A/MHz) + F_B\ (dB\ MHz) + T_{ij}(f_\ell)\ (dB) \\ + B\ (dB\ MHz) - I_s\ (dB\ \mu A),$$

where

$$M_{PP}^B(f_\ell) \qquad = 10\log m_{PP}^B(f_\ell)$$

$$Q\ (dB\ \mu A/MHz) = 20\log\left(\frac{\sqrt{2}q}{10^{-12}}\right)$$

$$T_{ij}(f_\ell)\ (dB) \qquad = 10\log t_{ij}(f_\ell)$$

$$B(dB\ MHz) \qquad = 20\log\left(\frac{b}{10^6}\right)$$

$$I_s(dB\ \mu A) \qquad = 20\log\left(\frac{I_r^s}{10^{-6}}\right).$$

3.4 Peak Current—Stochastic Waveforms

The peak value of a stochastic waveform cannot be given precisely. Therefore, the peak waveform susceptibility of a receptor must include, along with K, an estimate of the fraction of time that a stochastic waveform peak at the detector input must exceed K in order for interference to occur. This estimate is denoted as α. Let $i_r(t)$ be a stationary process which is adequately described by first- and second-order statistics (means and autocorrelations). Then the same holds for $i_d(t)$. For simplicity also assume $i_r(t)$ has zero mean. The $i_d(t)$ also has zero mean. The variance of $i_d(t)$ is given by

$$\sigma_d^2 = \int_{\infty}^{-\infty} G_r(f)|B_r(f)|^2 \, df. \tag{3.34}$$

Now, from Chebyshev's inequality the probability of $i_d(t)$ exceeding K is bound by

$$p(|i_d(t)| > k) < \frac{\sigma_d^2}{k^2}.$$

Define the susceptibility margin as

$$p(|i_d(t)| > k)/\alpha,$$

where α is the probability $(|i_d(t)| > k)$ which should not be exceeded. Then from before,

$$\frac{p(|i_d(t)| > k)}{\alpha} > \frac{\sigma_d^2/k^2}{\alpha}.$$

Therefore, an indication that a stochastic waveform is compatible, that is, does not cause interference, is given by

$$\frac{\sigma_d^2/k^2}{\alpha} < 1.$$

We now consider the computation at the input to the receptor. The variance of $i_r(t)$ is given by

$$\sigma_r^2 = \int_0^\infty G_r(f) \, df.$$

Then, using the following approximation for σ_d^2, we have

$$\sigma_d^2 < |B_r(f_p)|^2 \int_0^\infty G_r(f) \, df, \tag{3.35}$$

where f_p is the frequency for which $B_r(f)$ is maximum. Thus,

$$\sigma_d^2 < |B_r(f_p)|^2 \, \sigma_r^2.$$

Dividing the inequality by k^2 and after some manipulation, we arrive at the requirement for compatibility in the expression

$$\frac{\sigma_r^2}{\alpha |I_r^s(f_p)|^2} < 1,$$

and a peak current interference margin for stationary stochastic processes is given by

$$m_{PP}^B(f) = \frac{\sigma_r^2}{\alpha |I_r^s(f_p)|^2}.$$

The spectral level for a stochastic waveform transmitted by the jth emitter is given by

$$q = \left[\frac{P_T/(bw)}{(bw)r_{je}}\right]^{1/2},$$

and this expression is only for Gaussian noise. Following the identical steps outlined earlier for the peak current deterministic interference margin, the broadband point interference margin for a stochastic signal can be determined, and it follows that

$$m_{PP}^B(f) = \frac{q^2 (bw)t_{ij}(f)b}{\alpha |I_r^s(f_p)|^2}, \tag{3.36}$$

which is the broadband point interference margin for a stochastic signal. After conversion to decibels, the broadband point interference margin for a stochastic signal becomes

$$M_{PP}^B(f_c) \text{ (dB)} = Q \text{ (dB } \mu A/MHz) + BW \text{ (dB MHz)} + T_{ij}(f_c) \text{ (dB)} \\ + B \text{ (dB MHz)} - I_s \text{ (dB } \mu A) - \alpha \text{ (dB)},$$

where BW (dB MHz) $= 10 \log (bw/10^6)$, and α (dB) $= 10 \log (\alpha)$. The bw is the field intensity meter bandwidth.

The peak current interference margin for stationary waveforms (broadband Gaussian) is given by

$$I_{pc} = \frac{\text{erfc}\left(\dfrac{|I_r^s(f_p)|}{\sqrt{2}\sigma_r}\right)}{\alpha},$$

where erfc is the complementary error function,

$$\text{erfc}(u) = \frac{2}{\pi}\int_u^\infty e^{-t^2}\,dt.$$

Chapter 4 | Nonlinear Interference Models

4.0 Introduction

This section will provide detailed mathematical derivations of several nonlinear models. Since the basis of these derivations is the modified nonlinear transfer function, it is discussed in great detail in Section 4.1, with particular emphasis on its derivation from the more general Volterra series. After the general form of the nonlinear approach is developed, it will be used to derive the nonlinear models. This model will then be used to examine the limitations and approximations of the nonlinear transfer function approach to the Volterra analysis. The remainder of the chapter will then be devoted to the derivation of the remaining models, with particular emphasis on the assumptions used to obtain the models in a form suitable for a system-level analysis. This material is written with the help of Refs. [67–81].

4.1 The Modified Nonlinear Transfer Function Approach

4.1.1 THE VOLTERRA SERIES

The theory of functionals and functional expansions was first proposed by Vito Volterra in 1930. He established a working definition of a functional by noting that, just as a function operates on a set of variables to produce a new set of variables, a functional operates on a set of functions to produce a new set of functions. Using this definition, Volterra observed that an arbitrary functional could be expanded in what is now called a Volterra series, in a manner similar to the power series expansion of a function. He showed that every homogenous functional of degree n, acting on an arbitrary function, $x(t)$, could be written as

$$F_n[x(t)] = \int_a^b **** \int_a^b k_n(\zeta_1, \zeta_2, \ldots, \zeta_n) x(\zeta_1) \ldots (\zeta_n) \, d\zeta_1 \, d\zeta_2 \ldots d\zeta_n, \qquad (4.1)$$

where $[a,b]$ is the interval appropriate for the problem being considered.

Observing that the preceding equation holds, the Volterra series expansion of any arbitrary functional, $G[x(t)]$, may be written

$$G[x(t)] = \sum_{n=0}^{\infty} F_n[x(t)]$$

$$= k_0 + \int_a^b k_1(\zeta) x(\zeta) d\zeta + \int_a^b \int_a^b k_2(\zeta_1,\zeta_2) x(\zeta_1) x(\zeta_2) d\zeta_1 d\zeta_2 + \dots$$

(4.2)

The first important application of this Volterra series expansion to the analysis of nonlinear circuits was the relation of the output system, $y(t)$, to the input, $x(t)$, by a Volterra series of the form

$$y(t) = \sum_{n=1}^{\infty} y_n(t),$$

(4.3)

where the y_n are given by

$$y_n(t) = \int_{-\infty}^{\infty} \int_{-\infty}^{\infty} h_n(\tau_1 \dots \tau_n) x(t-\tau_1) \dots x(t-\tau_n) d\tau_1 \dots d\tau_n.$$

(4.4)

The simplification of Eq. (4.3) will provide the theoretical basis for our discussion of nonlinear interference effects. In analyzing this equation, Fourier transforms will be performed on various terms in the expansion, resulting in time- and frequency-domain representations of the input/output relationship. The inverse Fourier transform of the type

$$h_n(\tau_1 \dots \tau_n) = \int_{-\infty}^{\infty} \dots \int_{-\infty}^{\infty}$$

$$H_n(f_1, \dots, f_n) \exp\{i2\pi(f_1\tau_1 + \dots + f_n\tau_n)\} df_1 \dots df_n$$

(4.5)

will allow expression of Eq. (4.3) in terms of these $H_n(f)$. Therefore, if Eq. 4.5 is substituted into Eq. (4.3) and the convolutions over τ_k are performed, $y(t)$ is found to be

$$y(t) = \sum_{n=1}^{\infty} \int_{-\infty}^{\infty} \dots \int_{-\infty}^{\infty} H_n(f_1, \dots, f_n) X(f_1) X(f_2) \dots X(f_n)$$

$$\exp\{j2\pi(f_1 + f_2 + \dots + f_n)t\} df_1 \dots df_n.$$

By noting that the frequency spectrum of $y(t)$, $Y(f)$, is given by the output frequency spectrum,

$$Y(f) = \sum_{n=1}^{\infty} \int_{-\infty}^{\infty} \dots \int_{-\infty}^{\infty} H_n(f_1, \dots, f_n) X(f_1) X(f_2) \dots X(f_n) \int_{-\infty}^{\infty}$$

$$\exp\{-j2\pi(f - f_1 - f_2 \dots f_n)t\} dt\, df_1 \dots df_n.$$

The preceding equation may also be expressed in terms of the input/output frequency domain spectral relationship

$$Y(f) = \sum_{-\infty}^{\infty} \int_{-\infty}^{\infty} \cdots \int_{-\infty}^{\infty} H_n (f_1, \ldots, f_n) X_1 (f_1) \cdots X_n (f_n)$$
$$\times \, \delta (f - f_1 \cdots -f_n) \, df_1 \cdots df_n. \tag{4.6}$$

The relationships just given will be used to develop models which describe system degradation due to equipment nonlinearities. They are the time- and frequency-domain Volterra series which relate system output to various-order inputs.

The exact relationship between the Volterra and power series can be derived. In fact, it can be shown that the power series, representing a nonlinear system with no memory, is a special case of the more general Volterra analysis. Actually, the results yield

$$y(t) = \sum_{n=1}^{\infty} \int_{-\infty}^{\infty} \cdots \int a_n X(f_1) X(f_2) \cdots X(f_n)$$
$$\exp \{j2\pi(f_1 + f_2 + \ldots + f_n)t\} df_1 \cdots df_n, \tag{4.7}$$

which by definition of multidimensional Fourier transforms reduces to

$$y(t) = \sum_{n=1}^{\infty} a_n x^n(t). \tag{4.8}$$

The preceding two equations show that the Volterra series does reduce to the power series for a zero memory system, which helps explain why classical power series yields accurate results in cases with zero memory nonlinearities.

4.1.2 THE NONLINEAR TRANSFER FUNCTION APPROACH

The first simplification is that the system in question is only "mildly" nonlinear. Mildly nonlinear is, of course, an arbitrary defined concept, but will be utilized here to describe a nonlinear system which is characterized by only the first few terms of Eq. (4.3). The number of terms which must be retained is determined by the rate of convergence of Eq. (4.3). Thus, if only terms of degree $n < N$ are retained, input signals will be limited in amplitude to those which allow convergence of Eq. (4.3) within the first N terms. The second assumption made in this section is that inputs to the system of interest are sinusoidal. This appears to be a severe restriction, placing limitations on the applicability of Eq. (4.3) to phenomena which are not sinusoidal (Gaussian noise, etc.).

It has been shown, however, that Eq. (4.3) is valid for completely arbitrary inputs, and the equations derived using sinusoidal inputs have also been validated for arbitrary mild nonlinearities. This restriction is thus a legitimate approximation, which is valid for the types of inputs to be discussed in the sections which follow, and which leads to the equations from which the models are derived.

Using these assumptions, evaluation of Eq. (4.3) is straightforward, but cumbersome. Therefore, as the equation is simplified, the notation changes mentioned previously will be introduced to simplify bookkeeping and computational chores.

The first step in the derivation of the nonlinear transfer function series is to limit system nonlinearities so that terms of degree $n > N$ contribute negligibly to the response, $y(t)$. Equation (4.3) is thus written

$$y(t) = y_1(t) + y_2(t) + \ldots + y_N(t) = \sum_{n=1}^{N} y_n(t). \tag{4.9}$$

This situation, where the system is represented as N independent blocks, each having the common input $x(t)$, is depicted pictorially in Figure 4.1. The nonlinear transfer function approach is thus seen to represent the total response of a nonlinear circuit as the sum of N individual responses. The first-order response is character-

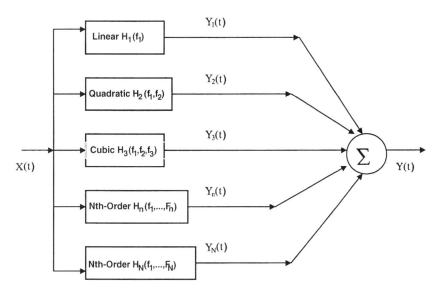

Figure 4.1 Model of weakly nonlinear circuit used by the nonlinear transfer function approach.

ized by the first-order linear transfer function $H_1(f_1)$, the second-order response is characterized by the second-order nonlinear transfer function $H_2(f_1,f_2)$, and higher-order responses are characterized by similar higher-order transfer functions.

4.1.2.1 Sinusoidal Steady-State Response of a Weakly Nonlinear System

To evaluate Equation (4.9), the second simplifying assumption will be utilized, and the input to the system will be represented as the sum of Q sinusoids,

$$x(t) = \sum_{q=1}^{Q} E_q \cos(2\pi f_q t), \tag{4.10}$$

where E_q is complex. If we let $E_q = E_{-q}$, $E = 0$, and $f_q = f_{-q}$ and note that

$$\cos x = \frac{e^{jx} + e^{-jx}}{2}$$

then $x(t)$ may be written in the complex plane as

$$x(t) = \frac{1}{2} \sum_{q=-Q}^{Q} E_q \exp(j2\pi f_q t). \tag{4.11}$$

Utilizing this form for the system input, we obtain the form of the nth-order response term

$$\begin{aligned} y_n(t) = \frac{1}{2^n} \sum_{q_1=-Q}^{Q} \cdots \sum_{q_n=-Q}^{Q} E_{q1} \ldots E_{qn} H_n(f_{q1} \ldots f_{qn}) \\ \exp\{j2\pi(f_{q1} + \ldots + f_{qn})t\}. \end{aligned} \tag{4.12}$$

This result shows that the application of a sum of Q sinusoids to a mildly nonlinear system yields additional output frequencies generated by the nth-order portion of the circuit. These additional output frequencies consist of all possible combinations of the input frequencies $f_{-Q}, \ldots, f_Q$, taken n at a time. At this point, a nonlinear change is made and the vector m is introduced to describe a particular frequency mix:

$$m = (m_{-Q}, \ldots m_{-1}, m_1, \ldots, m_Q). \tag{4.13}$$

This vector will represent all possible frequency mixes, since an individual m, say m_k, is defined as the number of times f_k appears in a particular frequency mix. The response frequency described by the m vector is thus

$$f_m = \sum_{k=-Q}^{Q} m_k f_k = (m_1 - m_{-1})f_1 + \ldots + (m_Q - m_{-Q})f_Q. \qquad (4.14)$$

For an nth-order portion of the response, the m_k's are constrained such that

$$\sum_{k=-Q}^{Q} m_k = n.$$

This, therefore, restricts the possible frequency mixes which may appear in the preceding equation quantizing all possible mixes.

4.1.2.2 Two-Tone Sinusoidal Response of a Weakly Nonlinear System

Combinatorial analysis yields the result that, if the excitation of Eq. (4.12) consists of Q sinusoids, the summation in (4.12) extends over

$$M = \frac{(2Q + n - 1)}{n!(2Q - 1)!}$$

distinct m vectors. Therefore, for a two-tone input (i.e., $Q = 2$),

$$M = \frac{(4 + n - 1)!}{n!(6)}.$$

Then $y_1(t)$ contains 4 frequency mixes, $y_2(t)$ contains 10 mixes, $y_3(t)$ contains 20 mixes, and so on.

Consider, then, a system where terms with $n > 3$ contribute negligibly to the output. There will then be 34 different frequency mixes arising from an input of the form

$$x(t) = \overline{E}_1 \cos 2\pi f_1 t + \overline{E}_2 \cos 2\pi f_2 t, \qquad (4.15)$$

which may be rewritten in the form

$$x(t) = \tfrac{1}{2}\{E_2 \,{}^*e^{-j2\pi f_2 t} + \overline{E}_1 \,{}^*e^{-j2\pi f_1 t} + \overline{E}_1 \, e^{j2\pi f_1 t} + \overline{E}_2 \, e^{j2\pi f_2 t}\}. \qquad (4.16)$$

Substituting Eq. (4.16) into Eq. (3.12) yields Table 4.1 which gives the 34 responses which must be summed to obtain $y_n(t)$.

Table 4.1   **First- and Second-Order Nonlinear Responses**

Combination No.	Combination				Frequency of Response	Amplitude of Response	Type of Response
	m_1	m_2	m_{-1}	m_{-2}			
$n = 1$							
1	1	0	0	0	f_1	$1/2\ E_1 H_1(f_1)$	Linear
2	0	1	0	0	f_2	$1/2\ E_2 H_1(f_2)$	"
3	0	0	1	0	$-f_1$	$1/2\ E_1 H_1(f_1)$	"
4	0	0	0	1	$-f_2$	$1/2\ E_2 H_1(f_2)$	"
$n = 2$							
1	1	1	0	0	$f_1 + f_2$	$1/2\ E_1 E_2 H_2(f_1, f_2)$	2nd-order interpolation
2	0	1	1	0	$f_2 - f_1$	$1/2\ E_1 E_2 H_2(-f_1, f_2)$	"
3	0	0	1	1	$-f_1 - f_2$	$1/2\ E_1{}^* E_2{}^* H_2(-f_1, -f_2)$	"
4	1	0	0	1	$f_1 - f_2$	$1/2\ E_1 E_2{}^* H_2(f_1, -f_2)$	"
5	1	0	1	0	$f_1 - f_1 = 0$	$1/2\ |E_1|^2 H_2(-f_1, f_1)$	"
6	0	1	0	1	$f_2 - f_2 = 0$	$1/2\ |E_2|^2 H_2(-f_2, f_2)$	"
7	2	0	0	0	$2f_1$	$1/4\ E_1^2 H_2(f_1, f_1)$	DC fixed
8	0	2	0	0	$2f_2$	$1/4\ E_1^2 H_2(f_2, f_2)$	"
9	0	0	2	0	$-2f_1$	$1/4\ E^*{}_1^2 H_2(-f_1, -f_1)$	2nd Harmonic
10	0	0	0	2	$-2f_2$	$1/4\ E^*{}_1^2 H_2(-f_2, -f_2)$	"
1	1	1	1	0	$f_1 + f_2 - f_1 = f_2$	$3/4\ E_1^2 E_2 H_3(f_1, f_2, -f_1)$	
2	0	1	1	1	$f_2 - f_1 - f_1 = -f_1$	$3/4\ E_1{}^* E_2 H_3(f_2, -f_1, -f_2)$	
3	1	0	1	1	$f_1 - f_1 - f_2 = -f_2$	$3/4\ E_1^2 E_2 H_3(f_1, -f_1, -f_2)$	
4	1	1	0	1	$f_1 + f_2 - f_2 = f_1$	$3/4\ E_1 E_2^2 H_3(f_1, f_2, -f_2)$	
5	2	1	0	0	$2f_1 + f_2$	$3/8\ E_1^2 E_2 H_3(f_1, f_1, f_2)$	
6	0	2	1	0	$2f_2 - f_1$	$3/8\ E_1{}^* E_2^2 H_3(f_1, f_2, -f_1)$	
7	0	0	2	1	$-2f_1 - f_2$	$3/8\ E^*{}_1^2 E^*{}_2 H_3(-f_1, -f_1, -f_2)$	
8	1	0	0	2	$f_1 - 2f_2$	$3/8\ E_1 E^*{}_2^2 H_3(f_1, -f_2, -f_2)$	
9	2	0	1	0	$2f_1 - f_1 = f_1$	$3/8\ E_1 E_1^2 H_3(f_1, f_1, -f_1)$	
10	0	2	0	1	$2f_2 - f_2 = f_2$	$3/8\ E_1 E_2^2 H_3(f_2, f_2, -f_2)$	
11	1	0	2	0	$f_1 - 2f_1 = -f_1$	$3/8\ E_1^2 E_2 H_3(f_1, -f_1, -f_2)$	
12	0	1	0	2	$f_2 - 2f_2 = -f_2$	$3/8\ E_1^2 E_2 H_3(f_2, -f_2, -f_2)$	
13	2	0	0	1	$2f_1 - f_2$	$3/8\ E_1^2 E^*{}_2 H_3(f_1, -f_1, -f_2)$	

Table 4.1   continued

Combination No.	Combination m_1 m_2 m_{-1} m_{-2}				Frequency of Response	Amplitude of Response	Type of Response
14	1	2	0	0	f_1+2f_2	$3/8\ E_1E_2^2H_3(f_1,f_2,f_2)$	2nd Harmonic
15	0	1	2	0	f_2-2f_1	$3/8\ E_2E_1^*H_3(f_2,-f_1,-f_1)$	"
16	0	0	1	2	$-f_1-2f_2$	$3/8\ E_1^*E_2^{*2}H_3(-f_1,-f_2,-f_2)$	"
17	3	0	0	0	$3f_1$	$1/8\ E_3^1H_3(f_1,f_1,f_1)$	"
18	0	3	0	0	$3f_2$	$1/8\ E_2^3H_3(f_2,f_2,f_2)$	"
19	0	0	3	0	$-3f_1$	$1/8\ E_1^{*3}H_3(f_2,f_2,f_2)$	"
20	0	0	0	3	$-3f_2$	$1/8\ E_2^{*3}H_3(-f_2,-f_2,-f_2)$	"

Each of these 34 frequency mixes represents a different nonlinear response: harmonic generation, intermodulation, desensitization, etc., as seen in the table. These 34 responses will be utilized in the next section to illustrate the final modification leading to the series which describes system-level nonlinear behavior.

4.1.3   THE MODIFIED NONLINEAR TRANSFER FUNCTION APPROACH

As a prelude to the final simplification to Eq. (4.9), consider the fact that the nonlinear transfer functions are, in general, complex functions which may be written

$$H_n\ (f_1,f_2,\ldots,f_n) = \left|H_n\ (f_1,f_2,\ldots,f_n)\right|e^{j\phi_n(f_1,f_2,\ldots)}, \qquad (4.17)$$

where ϕ_n is an arbitrary phase in the complex plane.

Consider also that the 34 responses in Table 4.1 occur at considerably fewer than 34 frequencies. The total response at each frequency is thus found by adding all individual responses at that frequency in the complex plane. This process can be illustrated for a particular case if the total $n < 3$ response at frequency f_1 is considered. The responses which must be summed to obtain the total response may be obtained from Table 4.1 and are as follows: for $n = 1$, the first response, and for $n = 3$, the fourth and ninth responses. Combining these in the complex plane results in Figure 4.2, which is a phasor diagram showing how the responses are added vectorially to obtain the total response at f_1, $y(t,f_1)$.

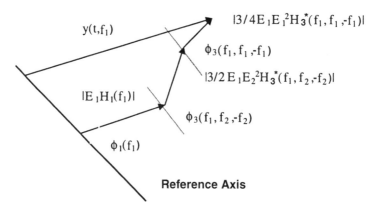

Figure 4.2   Phasor diagram of response at frequency f_1.

The final simplification of Equation (4.9) will involve limiting the phase of the nonlinear transfer function to either 0 or π. This is equivalent to considering the nonlinear transfer functions to be real functions, as opposed to the complex functions of the Volterra analysis. These functions, which will be called modified nonlinear transfer functions because of their derivations from the complex Volterra functions, are the functions used to describe system nonlinearities, where phase information is generally unavailable. The effect of limiting the transfer function in this manner will be examined in detail as models describing each of the individual effects are developed. The first effect to be considered is desensitization, which is examined in the next section.

4.2   Desensitization

Desensitization occurs when an interfering signal enters a receiver with sufficient magnitude to cause the receiver amplifiers to operate nonlinearly. This results in the response to the desired signal being ''desensitized'' because of the nonlinear operation of the amplifier. Desensitization can be a serious problem in a complex electromagnetic environment because its effects are cumulative; that is, all signals entering the receiver RF passband contribute to desensitization, and the resulting nonlinear operation may cause system degradation even if individual interfering signals cause no problems.

To consider the effects of a third-order desensitization, assume that the system of interest is only mildly nonlinear, and that terms of order $n > 3$ need not be considered. Assume also that inputs to the system are voltages, one of which is the desired signal and modulation,

$$S_i(t) = S[1 + s(t)] \cos \omega_s t, \tag{4.18}$$

while the other is an interfering unmodulated carrier,

$$I_i(t) = I \cos \omega_i t. \tag{4.19}$$

Assuming that noise is negligible, the total input to the system is simply the sum of the two signals $S_i(t) + I_i(t)$. This is a two-tone cosinusoidal input, and therefore the input/output voltage relationship may be derived from Table 4.1. Because of the presence of $s(t)$, it is assumed the system behaves quasi-statically. Similar assumptions will be made in the discussion of other nonlinear phenomena.

In the equation

$$V_o(t) = [1 + s(t)]SH_1(f_s) \cos \omega_s t + (3/2)I^2 SH_3(f_s, f_i, -f_i) \\ \times [1 + s(t)] \cos \omega_s t, \tag{4.20}$$

it can be seen that the cosine term involving ω_i has been eliminated. This arises from the factor of 1/2 in the second term of Eq. (4.20). The equation is actually of the form

$$V_o(t) = [1 + s(t)]SH_1(f_s) \cos \omega_s t + 3I^2 SH_3(f_s, f_i, -f_i) \\ \times [1 + s(t)] \cos \omega_s t \cos^2 \omega_i t. \tag{4.21}$$

However, since $\cos^2 \omega_i t = 1/2 + 1/2 \cos 2\omega_i t$, and since terms involving $\cos 2\omega_i t$ are eliminated due to receiver selectivity, Eq. (4.20) correctly describes the input/output relationship. Now, collecting similar terms in Eq. (4.20) yields

$$V_o(t) = S[H_1(f_s) + (3/2)I^2 H_3(f_s, f_i, -f_i)] * [1 + s(t)] \cos \omega_s t. \tag{4.22}$$

This equation represents the transfer function as a complex function with arbitrary phase. This relationship is depicted graphically in Figure 4.3. To obtain Eq. (4.22) in a form useful for a system-level analysis requires that these arbitrary phase angles be specified. This specification will eliminate phase considerations and involves two assumptions:

1. The linear (desired) portion of the response is entirely positive real (i.e., $\phi_1 = 0$).
2. The phase angle of H_3 will be limited to values of 0 or π.

If Eq. (4.22) is to represent desensitization, however, the actual system output will be less than the linear portion of the response. This leads to the requirement

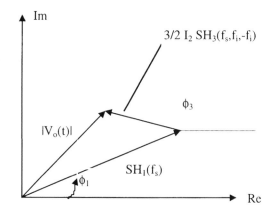

Figure 4.3   Phasor diagram of third-order desensitization.

that ϕ_3 should be approximated by π, which is equivalent to the statement that H_3, which is actually of the form $|H_3(f_s,f_i,-f_i)|e^{j\phi_3}$, is an entirely real, negative quantity, because $e^{j\pi} = -1$. Equation (4.22) may be rewritten using this requirement:

$$V_o(t) = S[H_1(f_s) - (3/2)I^2H_3(f_s,f_i,-f_i)] * [1 + s(t)] \cos \omega_s t. \quad (4.23)$$

Using Eq. (4.23), the effects of desensitization may be expressed as

$$\frac{\Delta S_o}{S_o} = \frac{S_o(\text{volts}) - S_o'(\text{volts})*}{S_o(\text{volts})} \quad (4.24)$$

where S_o (volts) is the desired signal output without interference, and S_o' (volts) is the desired signal output with interference:

$$\frac{\Delta S_o}{S_o} = \frac{3}{2} \frac{I^2H_3(f_s,f_s,-f_i)}{H_1(f_s)}.$$

The preceding equation can also be expressed in terms of decibels to give

$$\frac{\Delta S_o}{S_o} (\text{dB}) = 2P_1(\text{dB m}) + F(f_s,f_i) (\text{dBm}),$$

where

$$F(f_s, f_i) = -20 \log H_1/3H_3.$$

The preceding equations are for signals below the automatic gain control threshold (P_{AGC}), and for interfering signal power less than the saturation power level corresponding to the desired signal level and the frequency separation between the input signals. When the desired signal is above P_{AGC}, the gain is reduced proportionally to the increase in the desired signal so that the output remains constant. The gain reduction may be represented by

$$\Delta G \text{ (dB)} = k(P_D - P_{AGC}),$$

where

k = gain reduction fraction, assumed to be 1 if all AGC is applied prior to the nonlinearity

P_D = desired signal level in dBm, and P_{AGC} = AGC threshold in dBm.

The resulting equation for $\Delta S_o/S_o$ will be

$$\frac{\Delta S_o}{S_o} = 2P_1 \text{ (dBm)} - 2(P_D - P_{AGC}) + F(f_s, f_i) \text{ (dBm)}. \qquad (4.25)$$

Equation (4.25) is valid if only terms of order $N \leq 3$ must be considered to represent the transfer function of the nonlinear device. However, for any more than slight desensitization, higher-order terms must be considered. Higher-order terms could be added, but a series to represent the required circuitry is slowly converging and computation of the coefficients is, in general, not practical. The effects of considering only third-order desensitization may be examined by considering Figure 4.4, which shows the effect of phase angle on desensitization. This figure shows that higher-order terms and phase must be considered to accurately predict large desensitization. In fact, it has been shown that the equations given earlier are valid only for desensitization of approximately 1 dB.

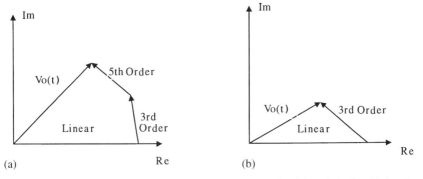

Figure 4.4 Effect of phase on desensitization: (a) actual (high-order); (b) third-order.

One might also represent the device as an ideal limiter, that is, constant gain for input signals below a saturation threshold, and complete saturation thereafter. By using a Fourier series (which allows more rapid convergence) and numerical integration to obtain the coefficient, the input–output relation can be computed. The resulting desensitization (S'_o/S_o) would be

$$D = \frac{2}{\pi} \arcsin \left(\frac{I_{sat}}{I} \right)$$

for $I \geq I_{sat}$, where I_{sat} is the saturation threshold in volts, and I is the interfering signal level in volts. This equation describes the desensitization of the desired signal as a function of the interfering signal level.

The desensitization may be expressed in decibels as

$$D \ (dB) = 20 \log [2/\pi \arcsin [\log^{-1}(P_{sat} - P_I)/20]], \tag{4.26}$$

where P_{sat} is the saturation power in dBm and P_I is the interfering signal power in dBm.

4.3   Gain Compression and Gain Expansion

The phenomenon of gain compression/expansion is very similar to that of desensitization. It, too, is a third-order effect which saturates the receiver amplifier stages and causes nonlinear operation. However, while desensitization is caused by an interfering signal, gain compression/expansion is caused by the desired signal, which may be of sufficient magnitude to cause nonlinear amplifier operation.

The equations describing gain compression/expansion are very similar to those presented in the last section. From Table 4.1, note that compression/expansion is obtained from $n = 1$, combination 1, and from $n = 3$, combination 9. It will be assumed that the desired signal is equivalent to Eq. (4.18) and that the other assumptions are as described in Section 4.2. The input–output equation may then be written

$$V_o(t) = S[H_1(f_s) \pm \tfrac{3}{4} S^2 H_3(f_s, -f_s, f_s)] [1 + s(t)]\cos \omega_s t. \tag{4.27}$$

In this equation, the $(+)$ and $(-)$ arise from phase considerations similar to those presented in Section 4.2 and refer to gain expansion and compression, respectively. This corresponds to considering gain expansion to have a phase of $\phi_3 = 0$ and gain compression to have phase of $\phi_3 = \pi$. This situation, described by Eq.

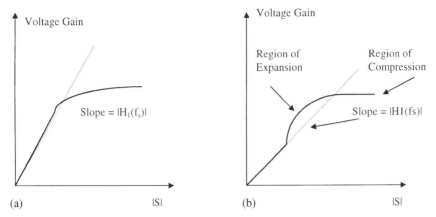

Figure 4.5   Voltage gain curves illustrating (a) gain compression and (b) both gain expansion and gain compression.

(4.27), where the voltage gain is not linear, but varies with desired signal power, may be seen in Figure 4.5.

At this point it becomes advantageous to consider only gain compression, since the discussion of gain expansion will be exactly the same, except for the (+) sign. This will simplify the discussion, while the results obtained will be easily adapted to describe expansion effects.

Consider the ratio

$$\frac{\Delta S_o}{S_o} = \frac{S_o(\text{volts}) - S'_o(\text{volts})}{S_o(\text{volts})} = \frac{3}{4} S^2 \frac{H_3(f_s, f_s, -f_s)}{H_1(f_s)} \tag{4.28a}$$

and

$$\frac{\Delta S_o}{S_o} (\text{dB}) = 20 \log S^2 + 20 \log \left| \frac{3}{2} \frac{H_3(f_s, f_s, -f_s)}{H_1(f_s)} \right| \tag{4.28b}$$
$$= 2P_s (\text{dBm}) + F(f_s) (\text{dBm}).$$

4.4   Intermodulation

In this section, models used to describe nonlinear intermodulation effects will be discussed. Intermodulation is the process occurring when two or more signals mix in a nonlinear device to produce an output at a frequency which causes performance degradation. Effects which will be discussed in this section are second-, third-, and fifth-order, two-signal intermodulation products, and three-signal, third-order intermodulation products. These effects will be considered for three cases: (1) intermodulation products generated in a receiver, (2) intermodulation products generated in a transmitter, and (3) intermodulation products generated in a nonlinear metallic junction.

These will be considered separately because there are several simplifications which make the examination of transmitter and receiver intermods much less involved than an examination of products generated in a metallic structure. The discussion of two-signal, third-order intermodulation will be quite detailed, whereas the equations describing higher-order effects will be extrapolated from the third-order, for although a rigorous mathematical derivation of these effects has been performed, to repeat it would merely lend complexity to the discussion without offering any additional insights. The models used to describe transmitter intermodulation products will be developed following the Volterra analysis of receiver intermods. Following the development of the preceding effects, the models utilized to describe structurally generated intermods will be presented. These equations will be based on empirical data, since structures cannot be considered mildly nonlinear, which prohibits descripton of their effects by a Volterra analysis.

4.4.1   TWO-SIGNAL, THIRD-ORDER RECEIVER
INTERMODULATION PRODUCTS

The assumptions used in the analysis of third-order intermodulation are the same as those used in the desensitization and gain compression/expansion analyses, with the exception of the representation used for the interfering signal. In this section, it is assumed that there are two interfering signal components, $I_1(t)$, and $I_2(t)$, present at the input of the nonlinear device. The first interfering signal component is assumed to be an unmodulated carrier and the second is amplitude modulated. The total interfering signal is thus represented by

$$I_i(t) = I_1 \cos \omega_1 t + I_2(1 + i(t))\cos \omega_2 t. \tag{4.29}$$

For this discussion of third-order intermodulation, the terms in the nonlinear transfer function expansion which must be considered are of the form

$$V_0(t) = S_i(t) + 3S_i(t)I_i^2(t) + I_i^3(t). \tag{4.30}$$

The nonlinear transfer functions have not yet been included in Eq. (4.30), although they are an inherent part of the Volterra analysis. They will be included following several simplifying assumptions. This will serve to simplify the notation considerably, with no loss of generality. The assumptions are as follows:

1. As has been discussed previously, the $S_i(t)I_i^2(t)$ term contributes to desensitization.
2. The signal at f_1 is assumed to be nearer the center of the receiver passband than the signal at f_2.
3. Of the possible third-order responses:
   (a) $2f_1 + f_2$
   (b) $2f_1 - f_2$
   (c) $2f_2 + f_1$
   (d) $2f_2 - f_1$

   the response (b) at $2f_1 - f_2$ is assumed to be the major interfering signal component. It is assumed that the others will be sufficiently attenuated by RF selectivity to be insignificant, since they fall farther from the center of RF passband than does (b).

Given assumption 1, Eq. (4.30) may be rewritten

$$V_o(t) = S_i(t) + I_i^3(t).$$

But $I_i^3(t)$ may be expanded using the binomial theorem:

$$I_i^3(t) = I_1^3(t) + 3I_1^2(t)I_2(t) + 3I_1(t)I_2^2(t) + I_2^3(t). \tag{4.31}$$

Now, making use of assumptions 2 and 3 leads to the input–output relationship

$$V_o(t) = S_i(t) + 3I_1^2(t)I_2(t). \tag{4.32}$$

Given that the interfering signal is represented by Eq. (4.29), Eq. (4.32) becomes

$$V_o(t) = S_i(t) + 3[I_1^2 \cos^2\omega_1 t \ I_2[1 + i(t)] \cos \omega_2 t]. \tag{4.33}$$

After some mathematical manipulation, we obtain

$$V_o(t) = SH_1(f_s)[1 + s(t)] \cos \omega_s t \tag{4.34}$$
$$+ \left(\tfrac{3}{4}\right)I_1^2 I_2 H_3(f_1, f_1, -f_2)[1 + i(t)] \cos(2\omega_1 - \omega_2)t.$$

The second term in Eq. (4.34) is the third-order intermodulation term.

Returning to Eq. (4.34), it may be seen that the amplitude of the third-order intermodulation carrier is given by

$$\text{IMV}_o(t) = \left(\tfrac{3}{4}\right)I_1^2 I_2 H_3(f_1, f_1, -f_2),$$

while the intermodulation output power in dBm is given by

$$P_{\text{IM}} = 2P_1 \text{ (dBm)} + P_2 \text{ (dBm)} + 20 \log (3/\sqrt{2}) + H_3(f_1,f_1,-f_2) \text{ (dB)}. \qquad (4.35)$$

For receivers, it is convenient to express the results in terms of an equivalent input power level that is required to produce the same effects in the receiver as the intermodulation product. The output for the desired signal may be found from Eq. (4.34) and is just the magnitude of the desired signal carrier:

$$\text{Desired signal output power (dBm)} = P_D \text{ (dBm)} + H_1(f_s) \text{ (dB)}. \qquad (4.36)$$

If this desired output power is assumed equal to the intermodulation output power, Eqs. (4.35) and (4.36) may be equated, resulting in

$$\begin{aligned}
P_L \text{ (dBm)} &= 2P_1 \text{ (dBm)} + P_2 \text{ (dBm)} \\
&\quad + 20 \log (3/\sqrt{2}) + H_3(f_1,f_1,-f_2) - H_1(f_s) \text{ (dB)} \qquad (4.37) \\
&= 2P_1 + P_2 + \text{IMF}(f_s,f_i,f_2) \text{ (dBm)},
\end{aligned}$$

where $\text{IMF}(f_s,f_1,f_2)$ is the intermodulation functional $-20 \log (\sqrt{2/3})$ (H_1/H_3) (dBW).

With reference to Eq. (4.37), the equivalent input signal for intermodulation is a function of the power levels of the two interfering signals, the nonlinearity factor (20 log $3/\sqrt{2} = k_3$), and the transfer functional. The problem becomes how to evaluate the intermodulation functional [20 log $3/\sqrt{2} + H_3(f_1,f_1,-f_2)$ $- H_1(f_s)$] for a particular receiver. As was the case with desensitization, it will be convenient to use specific data to evaluate the functional for a particular set of input conditions.

Intermodulation measurements made in accordance with MIL-STD are performed in a manner such that the equivalent intermodulation signal is equivalent to the receiver sensitivity, P_R (i.e., $P_D = P_R$) and the two interfering signals are equal in amplitude. If this is the case, and if $P_3(f_1,f_2)$ is defined as the power required for the signals at f_1 and f_2 to produce a standard response, then

$$\text{IMF}(f_s,f_1,f_2) = P_R - 3P_3^*(f_1,f_2). \qquad (4.38)$$

Substituting Eq. (4.38) into Eq. (4.37) yields

$$P_{\text{IM}} \text{ (dBm)} = 2P_1 \text{ (dBm)} + P_2 \text{ (dBm)} + P_R \text{ (dBm)} - 3P_3(f_1,f_2) \text{ (dBm)}. \qquad (4.39)$$

Equation (4.39) applies to intermodulation situations where the two signals producing the intermodulation do not saturate the receiver front end, and the resulting intermodulation does not exceed the receiver automatic gain control threshold. The next problem is to define what happens to the intermodulation product as the input signals are changed to levels which result in conditions other than those for which the spectrum signature measurements were performed.

As the input power for one or both of the intermodulation signals or the desired signal is increased, the resulting signal exceeds the receiver automatic gain control threshold, the receiver AGC is activated, and the receiver RF gain is reduced. For this situation, Equation (4.39) must be modified to account for the gain change resulting from the AGC:

$$P_{IM} \text{ (dBm)} = 2P_1 \text{ (dBm)} + P_2 \text{ (dBm)} + P_R \text{ (dBm)}$$
$$- 3P_3^*(f_1, f_2) \text{ (dBm)} + \Delta G \text{ (dB)}.$$

In addition, as either of the interfering signals is increased, a saturation level $P_{sat}(f)$ is reached such that additional increases in the interfering signal do not result in increases in the equivalent intermodulation input power. For this condition, the equivalent intermodulation input power may be represented as follows:

For $P_1(f_1) > P_{sat}(f_1)$:

$$P_{IM} \text{ (dBm)} = 2P_{sat}(f_1) \text{ (dBm)} + P_2 \text{ (dBm)} + P_R \text{ (dBm)} - 3P_3^*(f_1, f_2) \text{ (dBm)}.$$

For $P_2(f_2) > P_{sat}(f_2)$:

$$P_{IM} \text{ (dBm)} = 2P_1 \text{ (dBm)} + P_{sat}(f_2) + P_R \text{ (dBm)} - 3P_3^*(f_1, f_2) \text{ (dBm)}.$$

Table 4.2 summarizes the third-order intermodulation equations for the various conditions of interest.

Table 4.2 **Third-Order Intermodulation Equations**

Case	Condition	Equation
I	$P_1, P_2 < P_{sat}$ $P_{IM} < P_{AGC}$	$P_{IM} = 2P_1 + 2P_2 + \text{IMF}(f_1, f_2)$
II	$P_1, P_2 < P_{sat}$ $P_{IM} > P_{AGC}$	$P_{IM} = 2P_1 + 2P_2 + \text{IMF}(f_1, f_2) + \Delta G$
III	$P_1 > P_{sat}$ $P_2 < P_{sat}$	$P_{IM} = 2P_{sat} + 2P_2 + \text{IMF}(f_1, f_2)$
IV	$P_1 < P_{sat}$ $P_2 > P_{sat}$	$P_{IM} = 2P_1 + P_{sat} + \text{IMF}(f_1, f_2)$

$\text{IMF}(f_1, f_2) = P_R - 3P_3^*(f_1, f_2).$

4.4.2 SECOND- AND FIFTH-ORDER, TWO-SIGNAL RECEIVER INTERMODULATION PRODUCTS

As a continuation of the analysis presented in the previous section, consider a second-order intermodulation product occurring at frequency $f_1 \pm f_2$. After some manipulations it can be shown that

$$
\begin{aligned}
V_o(t) = S_i(t) &+ [1 + i(t)]I_1 I_2 [H_2(f_1,f_2)\cos(\omega_1 + \omega_2)t \\
&+ H_2(f_1,-f_2)\cos(\omega_1 - \omega_2)t].
\end{aligned}
\tag{4.40}
$$

Of course, only one of the frequencies, $f_1 + f_2$ or $f_1 - f_2$, will fall into the receiver passband, so Eq. (4.40) will reduce to either of two equations, corresponding to the $(+)$ and $(-)$ terms in

$$
\begin{aligned}
V_o(t) = SH_1(f_s) &[1 + s(t)]\cos \omega_s t \\
&+ I_1 I_2 H_2(f_1,\pm f_2)[1 + i(t)]\cos(\omega_1 \pm \omega_2)t.
\end{aligned}
\tag{4.41}
$$

From Eq. (4.41), a representation for the intermodulation output power may be given:

$$
P_{IM} \text{ (dBm)} = P_1 \text{ (dBm)} + P_2 \text{ (dBm)} + H_2(f_1,\pm f_2) \text{ (dB)}.
\tag{4.42}
$$

If Eq. (4.42) is presented in terms of an equivalent input power, a relationship analogous to Eq. (4.37) may also be found:

$$
\begin{aligned}
P_{IM} \text{ (dBm)} &= P_1 \text{ (dBm)} + P_2 \text{ (dBm)} + H_2(f_1,\pm f_2) \text{ (dB)} - H_1(f_s) \text{ (dB)} + 20\log 2 \\
&= P_1 \text{ (dBm)} + P_2 \text{ (dBm)} + \text{IMF}(f_1,f_2) \text{ (dBm)}.
\end{aligned}
\tag{4.43}
$$

To evaluate the IMF, again assume that the interfering signal powers are equal to each other and to $P_2^*(f_1,f_2)$, the power required to create a standard response. If this is the case, and if P_R is the receiver sensitivity, the IMF is given by

$$
\text{IMF}(f_1,f_2) = P_R - 2P_2^*(f_1,f_2)
\tag{4.44}
$$

and Eq. (4.43) becomes

$$
P_{IM} \text{ (dBm)} = P_1 \text{ (dBm)} + P_2 \text{ (dBm)} + P_R \text{ (dBm)} - 2P_2^*(f_1,f_2) \text{ (dBm)}.
\tag{4.45}
$$

In a manner analogous to that presented in Section 4.4.1, Eq. (4.44) may be modified to account for the effects of automatic gain control and receiver saturation. If these modifications are performed, the equation in Table 4.3 results.

The interference margin as determined by the criterion described is given by

$$
\text{Interference margin} = P_1 \text{ (dBm)} + P_2 \text{ (dBm)} - 2P_R \text{ (dBm)} - 132 \text{ (dB)}.
\tag{4.46}
$$

Table 4.3 **Second-Order Intermodulation Equations**

Case	Condition	Equation
I	$P_1, P_2 < P_{sat}$ $P_{IM} < P_{AGC}$	$P_{IM} = P_1 + 2P_2 + IMF(f_1, f_2)$
II	$P_1, P_2 < P_{sat}$ $P_{IM} > P_{AGC}$	$P_{IM} = P_1 + P_2 + IMF(f_1, f_2) + \Delta G$
III	$P_1 > P_{sat}$ $P_2 < P_{sat}$	$P_{IM} = P_{sat} + P_2 + IMF(f_1, f_2)$
IV	$P_1 < P_{sat}$ $P_2 > P_{sat}$	$P_{IM} = P_1 + P_{sat} + IMF(f_1, f_2)$

$IMF(f_1, f_2) = P_R - 2P_2^*(f_1, f_2)$.

To examine the effects of fifth-order intermodulation, it will be useful to obtain the fifth-order contribution to the fifth-order effect. This fifth-order effect is assumed to be the only significant response. The fifth-order term is

$$5/8 \ H_5(f_1, f_1, f_1, -f_2, -f_2)I_1^3 \ I_2^2[1 + i(t)]^2 \ \cos(3\omega_1 - 2\omega_2)t. \qquad (4.47)$$

From this, it is a simple matter to obtain the intermodulation carrier amplitude and the intermodulation output power:

$$IMV_o(t) = 5/8 \ H_5(f_1, f_1, f_1, -f_2, -f_2)I_1^3 \ I_2^2$$

and

$$P_D \ (dBm) = 3P_1 \ (dBm) + 2P_2 \ (dBm) + 20 \log (5/8) \qquad (4.48)$$
$$+ \ 20 \log 4\sqrt{2} + H_5 \ (f_1, f_1, f_1, -f_2, -f_2)(dB).$$

Using the assumption regarding standard responses presented previously yields

$$P_D \ (dBm) = 3P_1 \ (dBm) + 2P_2 \ (dBm) + 2 - \log (5/8)$$
$$+ \ 20 \log 4\sqrt{2} + H_5(f_1, f_1, f_1, -f_2, -f_2) \ (dB) - H_1(f_s), \qquad (4.49)$$

whereas evaluation of the IMF leads to

$$IMF(f) = P_R - 5P_5^*(f_1, f_2). \qquad (4.50)$$

Substituting Eq. (4.43) into Eq. (4.49) the equivalent intermodulation power may be obtained:

$$P_{IM} \ (dBm) = 3P_1 \ (dBm) + 2P_2 \ (dBm) + P_R \ (dBm) - 5P_5^*(f_1, f_2). \qquad (4.51)$$

Equation (4.51) may be modified to account for the effect of AGC and saturation. The modifications are very similar to those performed to obtain Tables 4.2 and 4.3, and these changes lead to the results presented in Table 4.4. The default is found to be

$$P_{IM} \text{ (dBm)} = 3P_1 \text{ (dBm)} + 2P_2 \text{ (dBm)} - 5P_R \text{ (dBm)} - 330 \text{ dB}.$$

4.4.3 THREE-SIGNAL, THIRD-ORDER RECEIVER INTERMODULATION PRODUCTS

Equations similar to those in previous sections may be developed for third-order, three-signal intermods, where the input will be at frequencies of the form

$$F_{IM} = \pm f_1 \pm f_2 \pm f_3. \tag{4.52}$$

Because of RF selectivity, however, all three input signals must be approximately equal, or they will be attenuated and cause no degradation. Because of this, and because F_{IM} must also fall in the RF passband, Eq. (4.52) will be constrained to frequency mixes such that two of the three frequencies will be positive. For the purpose of the following mathematical discussions, the intermodulation frequency will be represented by

$$F_{IM} = f_1 + f_2 - f_3.$$

Based on an analysis similar to that presented in Section 4.4.1, the input/output relationship may then be expressed by

$$V_o(t) = \left(\tfrac{3}{2}\right) I_1 I_2 I_3 H_3(f_1, +f_2, -f_3)\cos(\omega_1 + \omega_2 - \omega_3)t. \tag{4.53}$$

Table 4.4 **Fifth-Order Intermodulation Equation**

Case	Condition	Equation
I	$P_1, P_2 < P_{sat}$ $P_{IM} < P_{AGC}$	$P_{IM} = 3P_1 + 2P_2 + \text{IMF}(f_1, f_2)$
II	$P_1, P_2 < P_{sat}$ $P_{IM} > P_{AGC}$	$P_{IM} = 3P_1 + 2P_2 + \text{IMF}(f_1, f_2) + \Delta G$
III	$P_1 > P_{sat}$ $P_2 < P_{sat}$	$P_{IM} = 3P_{sat} + 2P_2 + \text{IMF}(f_1, f_2)$
IV	$P_1 < P_{sat}$ $P_2 > P_{sat}$	$P_{IM} = 3P_1 + 2P_{sat} + \text{IMF}(f_1, f_2)$

$\text{IMF}(f_1, f_2) = P_R - 5P_5^*(f_1, f_2).$

From this, it is possible to find the intermodulation output power:

$$P_{IM} \text{ (dBm)} = P_1 \text{ (dBm)} + P_2 \text{ (dBm)} + P_3 \text{ (dBm)} + 20 \log (12)$$
$$+ H_3(f_1,f_2,-f_3) \text{ (dB)}.$$

Equating the intermodulation output power and the equivalent desired power leads to

$$P_{IM} \text{ (dBm)} = P_1 \text{ (dBm)} + P_2 \text{ (dBm)} + P_3 \text{ (dBm)} + 20 \log (12)$$
$$+ H_3(f_1,f_2,-f_3) \text{ (dB)} - H_1(f_s) \quad\quad (4.54)$$
$$= P_1 \text{ (dBm)} + P_2 \text{ (dBm)} + P_3 \text{ (dBm)} + IMF(f_1,f_2,f_3).$$

It will again be useful to evaluate the IMF in terms of specific data. Given the assumptions of those two equations, the IMF may be written as

$$IMF(f_1,f_2,f_3) = P_R - 3P_3{}^* (f_1,f_2,-f_3),$$

which yields

$$P_{IM} \text{ (dBm)} = P_1 \text{ (dBm)} + P_2 \text{ (dBm)} + P_3 \text{ (dBm)} \quad\quad (4.55)$$
$$- 3P_3{}^*(f_1,f_2,-f_3) \text{ (dBm)}.$$

The results of incorporating AGC and saturation effects into Eq. (4.55) lead to the equations of Table 4.5.

4.4.4 TRANSMITTER INTERMODULATION PRODUCTS

In addition to the intermodulation products generated in receivers discussed in previous sections, products may also be generated in the nonlinear portions of

Table 4.5   Third-Order, Three-Signal Intermodulation Equations

Case	Condition	Equation
I	$P_1,P_2,P_3 < P_{sat}$ $P_{IM} < P_{AGC}$	$P_{IM} = P_1 + P_2 + P_3 + IMF(f_1,f_2,f_3)$
II	$P_1,P_2,P_3 < P_{sat}$ $P_{IM} > P_{AGC}$	$P_{IM} = P_1 + P_2 + P_3 + IMF(f_1,f_2,f_3) + \Delta G$
III	$P_1 > P_{sat}$ $P_2,P_3 < P_{sat}$	$P_{IM} = P_{sat} + P_2 + P_3 + IMF(f_1,f_2,f_3)$
IV	$P_1,P_2 > P_{sat}$ $P_3 < P_{sat}$	$P_{IM} = 2P_{sat} + P_3 + IMF(f_1,f_2,f_3)$

$IMF(f_1,f_2) = P_R - 3P_3{}^*(f_1,f_2,-f_3).$

transmitters. Because of the nature of transmitter nonlinearities, however, Volterra techniques are not the appropriate analytical tool for use in the study of transmitter intermods. Instead, the intermodulation output power, which propagates from one of the interfering transmitters to the affected receiver, is of the form

$$P_{IM} \text{ (dBm)} = P_{low} \text{ (dBm)} - A \text{ (dB)} - B \log \Delta f\%, \qquad (4.56)$$

where P_{low} is the lowest interfering signal power, $\Delta f\%$ is the average percent difference of incoming transmitter frequencies from the mixing transmitter frequency, and A,B are the constants which must be determined for each transmitter and each product. Based on Eq. (4.56), default models may be formulated:

Third-order, two-signal or three-signal:

1. If $\Delta f\% \leq 1\%$: $A = 10$, $B = 0$, $P_{IM} = P_I$ (dBm) $- 10$ dB.
2. If $\Delta f\% > 1\%$: $A = 10$, $B = 30$, $P_{IM} = P_I$ (dBm) $- 30 \log \Delta f\%$ $- 10$ dB.

Fifth-order, two-signal:

1. If $\Delta f\% < 1\%$: $A = 30$, $B = 0$, $P_{IM} = P_I$ (dBm) $- 30$ dB.
2. If $\Delta f\% > 1\%$: $A = 30$, $B = 30$, $P_{IM} = P_I$ (dBm) $- 30 \log \Delta f\%$ $- 30$ dB.

Second-order, two-signal: Because of the nature of transmitter nonlinearities, second-order, two-signal intermodulation products are insignificant. This may be seen by considering the intermodulation output frequency $F_{IM} = f_1 \pm f_2$. It is obvious that at least one of the frequencies f_1, f_2, or f_{IM} will always be outside the transmitter operating band, which will cause the power at the out-of-band frequency to be attenuated to such an extent that the product is insignificant and causes no system degradation. For reference, the default models for all receiver and transmitter intermodulation products are presented in Table 4.6.

4.5   Cross Modulation

Cross modulation is the term used to describe degradation caused by the transfer of modulation from an interfering signal to the desired signal. Cross modulation is similar to desensitization (discussed earlier) in several ways which will bear on the following discussion: (1) Cross modulation is treated as a third-order effect, and (2) cross modulation may be considered a nonlinear phenomenon which occurs at intervals corresponding to increased interfering signal levels.

Table 4.6   **Intermod Default Models**

Order	Receiver	Transmitter
2(2 sig)	PIM $= P_1 + P_2 - P_R - 132$ dB	—
3(2 sig)	PIM $= 2P_1 + P_2 - 2P_R - 198$ dB	$P_{IM} = P_1 - 10$ dB *or* $P_{IM} = P_1 - 10$ dB $- 30 \log \Delta f\%$
3(3 sig)	PIM $= P_1 + P_2 + P_3 - 2P_R - 198$ dB	Same as above
5(2 sig)	PIM $= 3P_1 + 2P_2 - 4P_R - 330$ dB	PIM $= P_1 - 10$ dB *or* $P_{IM} = P_1 - 10$ dB $- 30 \log$ $\Delta f\%$

Based on the preceding similarities, the discussion of cross modulation will closely follow that of desensitization, except for the representation of the interfering signal, which is assumed to be an amplitude-modulated carrier, with modulation such that $i(t)$ is less than 1:

$$I_i(t) = I[1 + i(t)]\cos \omega_i t. \tag{4.57}$$

Then, one may obtain the input–output relationship

$$V_o(t) = S_i(t) + 3S_i(t)I_i^2(t). \tag{4.58}$$

Expanding the expression $I_i^2(t)$,

$$I_i^2(t) = I^2[1 + 2i(t) + i^2(t)] [1/2 + (1/2) \cos 2\omega_i t]. \tag{4.59}$$

Again, as in the case of the desensitization analysis, terms involving $\cos 2\omega_i t$ will be removed by filtering. Therefore, the significant terms in the output are

$$V_o(t) = S[H_1(f_s) + (3/2)I^2 H_3(f_s,-f_i, f_i) [1 + 2i(t)]] (1 + s(t)) \cos \omega_s t. \tag{4.60}$$

In Eq. (4.59), the term involving $i^2(t)$ is second-order and is insignificant with respect to the remainder of the expression. Now, carrying out the multiplications in Eq. (4.60), and eliminating terms with second-order modulation, the relation may be written

$$V_o(t) = S[[H_1(f_s) + (3/2)H_3(f_s,-f_i,f_i)I^2]$$
$$\times [1 + s(t)]H_3 (f_s, f_i, -f_i) [3I^2 i(t)]]\cos \omega_s t,$$

which may be written

$$V_0(t) = S\left[H_1(f_s) + \frac{3}{2}H_3(f_s,f_i,-f_i)I^2\right]$$

$$\times [1 + s(t) + \frac{(3/2)H_3(f_s,f_i,-f_i)I^2}{H_1(f_s) + (3/2)H_3(f_s,f_i,-f_i)I^2} \, 2i(t)]\cos \, \omega_s t. \quad (4.61)$$

Equation (4.61) is the expression for an amplitude-modulated signal where the modulation consists of a combination of the desired and interfering signal modulations, $s(t)$ and $i(t)$. If the modulation component resulting from the interfering signal modulation is restricted so that the maximum amplitude of the modulation signal is less than or equal to the amplitude of the carrier, overmodulation is avoided. In this case, the nonlinearity should not cause significant distortion of the modulation.

One measure of the effect of cross modulation is provided by the ratio of the sideband component resulting from the desired signal to the sideband component resulting from the interfering signal. The resulting ratio, which is termed the "cross modulation ratio," is expressed in terms of m_s and m_i (the maximum value of $s(t)$ and $i(t)$) as

$$\text{CMR} = \frac{[H_1(f_s) + (3/2)I^2 H_3(f_s,f_i,-f_i)]m_s}{[(3/2)I^2 H_3(f_s,f_i,-f_i)2m_i}. \quad (4.62)$$

Expressing the ratio in terms of m_s and m_i provides an effective measure when similar modulation is present on both the desired and interfering signals. If they are significantly different, it may be desirable to use some other measure of the amplitudes (such as the RMS levels).

If the desired and interfering signals are limited to "small signal" conditions such that

$$H_1(f_s) >> (3/2)I^2 H_3(f_s,f_i,-f_i). \quad (4.63)$$

The cross modulation ratio may then be written

$$\text{CMR} = \frac{H_1(f_s)m_s}{3I^2 H_3(f_s,f_i,-f_i)m_i} \quad (4.64)$$

or

$$\text{CMR (dB)} = -2P_I \text{ (dBm)} + H_1(f_s) - H_3(f_s,f_i,-f_i) \text{ (dB)} \\ - 20 \log \{2m_s/3m_i\}, \quad (4.65)$$

where P_I is the interfering signal power in dBm.

Equation (4.65) assumes that the major cross modulation is of third-order origin and results in an amplitude modulation characteristic (i.e., $m_s = m_i$). The last three terms of Eq. (4.65) are functions of the gain, selectivity, and nonlinear characteristics of the device under consideration. For convenience, these terms may be represented by a single functional, $CMF(f_I)$, which is referred to as the cross modulation function. Thus,

$$CMR \text{ (dB)} = -2P_I \text{ (dBm)} + CMF(f_i, f_s) \text{ (dBm)}. \qquad (4.66)$$

Equation (4.66) is valid for desired signals below the automatic gain control threshold (P_{AGC}) and for interfering signal powers less than the saturation power level corresponding to the desired signal level and the frequency separation between the input signals. When the desired signal is above P_{AGC}, the gain is reduced proportional to the increase in the desired signal so that the output remains constant. The gain reduction may be represented by

$$\Delta G \text{ (dB)} = k(P_D - P_{AGC}),$$

where k is the gain reduction fraction ($=1$ since all AGC occurs prior to nonlinearity), P_D is the desired signal level in dBm, and P_{AGC} is the AGC threshold in dBm. The resulting equation for the cross modulation ratio will be

$$CMR \text{ (dB)} = -2P_I \text{ (dBm)} - 2(P_D - P_{AGC}) + CMF(f_I) \text{ (dBm)}. \qquad (4.67)$$

When the interfering signal becomes large, the nonlinear device will saturate. Beyond this level, if the desired signal is constant, changes in the interfering signal level do not produce corresponding changes in the signal-to-interfering ratio. As a first approximation of the signal-to-interfering ratio P_I greater than the saturation level, P_{sat}, the saturation level may be substituted for the interfering signal level. Therefore,

$$CMR \text{ (dB)} = -2P_{sat} + 2(P_{AGC} - P_D) + CMF(f_s, f_i). \qquad (4.68)$$

The equations for cross modulation effects are summarized in Table 4.7.

In order to use these equations, it is necessary to specify a value for the cross modulation functional. If the cross modulation effects are specified for a given interfering signal level, $P_I^*(f_I)$ (dB), the cross modulation functional may be evaluated as

$$CMF(f_I) \text{ (dB)} = 2P_I^*(f_I) \text{ (dBm)} + CMR^* \text{ (dB)}, \qquad (4.69)$$

where CMR^* (dB) is the cross modulation ratio resulting from the reference interfering signal.

Table 4.7   **Summary of Equations for Cross Modulation**

Case	Conditions	Equation
I	$P_I < P_{sat}, P_D < P_{AGC}$	$CMR = -2P_I + CMF(f_s, f_i)$
II	$P_I < P_{sat}, P_D \geq P_{AGC}$	$CMR = -2P_I + 2(P_{AGC} - P_D)CMF(f_s, f_i)$
III	$P_I \geq P_{sat}, P_D \geq P_{AGC}$	$CMR = -2P_{sat} + 2(P_{AGC} - P_D) +$ $CMF(f_s, f_i)$
IV	$P_I \geq P_{sat}, P_D < P_{AGC}$	$CMR = -2P_I + CMF(f_s, f_i)$

From MIL-STD-461, the interfering signal power will be 66 dB above some standard reference, assumed to be the receiver sensitivity, so

$$P_I^*(f_1) \text{ (dBm)} = P_R \text{ (dBm)} + 66 \text{ dB} \tag{4.70}$$

and

$$CMR^* \text{ (dB)} = 0.$$

Then the CMF may be written

$$CMF^* \text{ (dB)} = 2P_R \text{ (dBm)} + 132 \text{ dB}, \tag{4.71a}$$

and the default model is

$$CMR \text{ (dB)} = -2P_I \text{ (dBm)} + 2P_R \text{ (dBm)} + 132 \text{ dB}. \tag{4.71b}$$

It is possible to define an interfering margin for cross modulation in much the same way the desensitization interference margin was defined. The interference margin will be defined as the amount by which the CMR exceeds a reference of -20 dB, the minimum observable cross modulation. Equation (4.71b) may then be written using this criterion:

$$\text{Interference margin} = CMR \text{ (dB)} + 20 \text{ dB}.$$

The equations just presented describe modulation of a desired AM signal by an interfering AM signal. Expressions similar to those in Table 4.7 can be derived for other types of cross modulation.

For interfering signals other than full-carrier, double-sideband AM, another approach is used to evaluate to cross modulation signal-to-interference ratio. This approach still correlates cross modulation with desensitization, but from a different point of view. Large signals entering a receiver cause desensitization, that is, gain reduction to the desired signal. If the bandwidth of the RF stages of a receiver is large enough, the gain reduction will follow the amplitude variations of the interfering signal and in this manner impart the unwanted modulation to the desired transmission.

Using this hypothesis, a modulation index, analogous to the AM modulation index, can be computed from the desensitization information. The expression for modulation index of an AM signal in terms of maximum and minimum instantaneous signal amplitude is

$$m_i = \frac{S_{max} - S_{min}}{S_{max} + S_{min}} = \frac{\Delta S_o / S_o}{2 - \Delta S_o / S_o}, \tag{4.72}$$

where $\Delta S_o / S_o$ is defined and where m_i is the modulated index, S_{max} is the maximum instantaneous signal amplitude, and S_{min} is the minimum instantaneous signal amplitude. Since the CW signal now has modulation on it, the signal-to-interference ratio is the ratio of the carrier amplitude to the sideband amplitude:

$$S/I = 20 \log (m_i). \tag{4.73}$$

Suppose instead of a CW carrier, the desired signal has been amplitude modulated. Then the sideband-to-sideband ratio would be

$$S/I = 20 \log (m_s / m_i), \tag{4.74}$$

where m_s is the desired signal modulation index, and m_i is the equivalent interfering signal index. Equation (4.74) is the cross modulation signal-to-interference model chosen for single-sideband interference to AM receivers, when m_i is computed by equation (4.72). The desired signal modulation index for AM signals is proportional to the total power in the modulation sidebands. Actually, the fraction of total power in the sidebands equals $m_s/2$. Equation (4.74) can be modified to describe single-sideband interference to a single-sideband receiver if one considers that for this type of desired signal, all power is contained in the information sidebands. Equation (4.75) expresses the cross modulation signal-to-interference ratio for SSB interference to SSB receivers:

$$S/I = 20 \log (2 / m_i). \tag{4.75}$$

Equation (4.75) could also model pulse interference to AM receivers, except that the interference appearing at the audio output of the receiver is proportional to the average power because, normally, the pulse bandwidth is considerably larger than the receiver's last IF passband. A bandwidth correction factor is also needed. The signal-to-interference ratio from cross modulation due to pulse interference is

$$S/I = 20 \log(m_s/m_i) - 10 \log (\tau f_r) - 10 \log (\tau \Delta f_3), \qquad (4.76)$$

where τ is the pulse width (sec), f_r is the pulse repetition frequency (pps), and Δf_3 is the receiver overall 3-dB bandwidth.

This can be seen if we examine the relationship of the S/I peak, given by Eq. (4.74), to S/I_{peak} minus the correction factors described earlier, since the factors affect the interference power, which is in the denominator of Eq. (4.74). Therefore,

$$S/I_{\text{avg}} = 20 \log m_s/m_i - 20 \log \tau/T - 10 \log \Delta f_3/f_r. \qquad (4.77a)$$

But since

$$20 \log \tau/T = 20 \log \tau f_r \qquad (4.77b)$$

and

$$10 \log \Delta f_3/f_r = 10 \log \tau \Delta f_3 - 10 \log \tau f_r, \qquad (4.78)$$

S/I_{avg} is then seen to be the expressions given by Eq. (4.76). This situation, where the interfering pulse width is much greater than the receiver bandwidth, is seen in Figure 4.6.

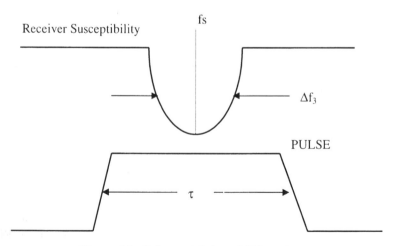

Figure 4.6 Pulse modulation of AM receivers.

By applying the same rationale that was employed to arrive at Eq. (4.74), from the expression of single-sideband interference to an AM receiver, an expression of pulse interference to a single-sideband receiver is given by

$$S/I = 20 \log (2/m_i) - 10 \log (\tau f_r) - 10 \log (\tau \Delta f_3). \qquad (4.79)$$

By applying the same logic used to obtain the form of the interference margin, all of the signal-to-interference ratios in this section may be expressed as interference margins using the relationship

$$\text{Interference margin} = S/I \text{ (dB)} + 20 \text{ dB}. \qquad (4.80)$$

This section has presented equations describing third-order cross modulation of many signal types. Three notable signal combinations were excluded from consideration, however, based on the following reasons:

1. PM cross modulation is not considered because the frequency information in an FM signal is not adversely affected by cross modulation amplitude variations. In addition, since most FM receivers have some sort of amplitude limiting, the modulation fluctuations would be undetected.

2. Pulse modulation of pulse signals is deemed insignificant because cross modulation between pulses depends on their simultaneous occurrence. Considering typical pulse duty cycles, this simultaneous occurrence is highly unlikely and was not considered in this analysis.

3. Cross modulation of pulse signals is deemed insignificant because it does not usually result in degradation of these types of receptors.

4.6 Spurious Responses

A spurious response in a superheterodyne receiver arises when an interfering signal (or one of its harmonics) enters a nonlinear mixer and combines with the local oscillator frequency of the mixer (or one of its harmonics) to produce a "spurious" output which falls into the receiver IF passband. This can be a serious problem because of the large amplitude of the local oscillator signal, which can mix with even small interfering amplitudes to produce system degradation. A diagram of a typical superheterodyne receiver, with three stages of nonlinear mixing, may be seen in Figure 4.7. For the first mixer in the figure, assume the interfering signal (or one of its harmonics, denoted by q) passes through RF preselection and amplification with sufficient power to enter mixer 1. When

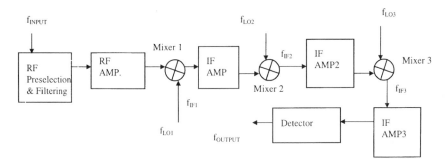

Figure 4.7   Three-stage superheterodyne receiver.

mixed with the first LO frequency (or one of its harmonics, denoted by p), the output frequency will occur at

$$f_{out} = pf_{LO} \pm f_{SPUR}. \tag{4.81}$$

In order to produce degradation, f_{out} must be within the first IF passband of the receiver. This will encompass a range of frequencies denoted by

$$f_{OUT} = f_{IF1} \pm \Delta f_{IF1} \tag{4.82}$$

Therefore, degradation will occur for a range of interfering frequencies such that

$$f_{SPUR} = p_1 f_{LO1} \pm (f_{IF1} \pm \Delta f_{IF1}) / q_1. \tag{4.83}$$

Now, given that Eq. (4.83) holds at each mixer, the requirement that f_{SPUR} causes interference in a triple conversion receiver leads to the requirement that

$$f_{SPUR} = \frac{p_1 f_{LO1}}{q_1} \pm \frac{p_2 f_{LO2}}{q_1 q_2} \pm \frac{p_3 f_{LO3} \pm (f_{IF3} \pm \Delta f_{IF3})}{q_1 q_2 q_3}. \tag{4.84}$$

From Eqs. (4.83) and (4.84), it is obvious that the interfering signal need not bear the center of the receiver RF passband to cause degradation and may, in fact, be greatly attenuated prior to mixing with the large-amplitude local oscillator frequency. This leads to two observations regarding spurious responses:

1. Since the spurious response frequencies need not be in the RF passband, only nonadjacent channel effects will be considered. This will introduce an additional cull at each interfering frequency.

2. Because of the large signal nature of the local oscillator signal (typically -1 V compared with an RF input to the mixer with amplitude on the

order of 1 mV), the modified Volterra analysis presented previously is not applicable to the study of spurious responses.

As a result of these observations, the algorithms used to describe spurious response interference will be considered differently from the algorithms used in the Volterra analysis. These differences will now be examined in more detail. Observation 1 states that the spurious responses need not be an adjacent-channel effect. If the interfering signal is nonadjacent channel, however, it is assumed that the RF attenuation is such that only first-mixer-generated responses will cause degradation. The second and third stages will be assumed to provide only direct IF feedthrough (i.e., normal mixing).

These two observations, coupled with the large signal local oscillator (allowing large values of p_1) will lead to nonadjacent channel spurious responses being qualified as follows:

$$\text{Mixer 1: } p_1 = \text{user-defined range}$$
$$q_1 = \text{user-defined or } q = 1 \text{ (default).}$$
$$\text{Mixer 2: } p_2 = q_2 = 1.$$
$$\text{Mixer 3: } p_3 = q_3 = 1.$$

If the interfering signal is adjacent channel, it will experience considerably less RF attenuation than a nonadjacent-channel signal. Therefore, p_1 and q_1 will be allowed to take on a range of values determined by the user. As may be seen in Figure 4.8, however, higher-order mixers require very high input signals to cause degradation. It is thus deemed advisable to limit the second and third mixers to values of p and q which are equal. As seen from Figure 4.8, this will result in an output near the IF frequency, which will pass relatively unattenuated to the next receiver stage. Using the preceding assumption yields the adjacent-channel response quantization:

$$\text{Mixer 1: } p_1 = \text{user-defined range}$$
$$q_1 = \text{user-defined range.}$$
$$\text{Mixer 2: } p_2 = q_2 = 1,2,3.$$
$$\text{Mixer 3: } p_3 = q_3 = 1,2,3.$$

Observation 2 will now be used to describe the response at the frequencies given earlier.

Because of the large signal nature of the local oscillator, an alternative to the Volterra analysis is needed to describe system degradation due to spurious responses. For $q = 1$, these response curves will be of the form

$$P_{SR} \text{ (dBm)} = I \log p + J. \tag{4.85}$$

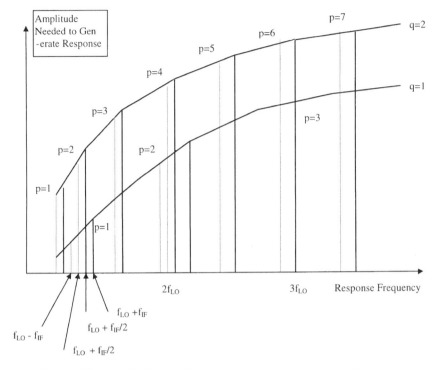

Figure 4.8   Amplitude needed to generate response vs response frequency.

In general, I and J must be determined for each particular receiver under consideration. In fact, it is often necessary to consider that the power needed to produce a response will have a different linear relationship in different receiver frequency intervals. This will necessitate determination of other values for I and J for additional frequency intervals in the receiver passband. Equation (4.85) will hold only in the nonadjacent-channel region. In the adjacent-channel region, the response will be of the form

$$P_{SR} \text{ (dBm)} = I \log f/f_o + J. \tag{4.86}$$

The spurious response powers may be related to the interference margin by noting that the interference margin is just the actual interfering signal minus P_{SR}.

As a rule of thumb, experimental data has shown that the $q = 2$ responses will be approximately 15 dB below the $q = 1$ responses, and the $q = 3, 4$ responses will be on the order of 20 dB below the $q = 1$ responses. The piecewise linear model described earlier has been validated and is discussed in detail in references

[68] and [79]. This model may be used to formulate default equations for spurious responses based on the CS04 limits of MIL-STD-461. The default will be formulated by assuming the response, P_{SR}, will be a signal of the form seen in Figure 4.6. This assumption will then be utilized to solve for I and J in the various regions of interest. The P_{SR} (in the region outside the 80-dB bandwidth) will be 15 dB lower if $q = 2$ and 20 dB lower if $q = 3$ or 4. Inside the 80-dB bandwidth region, the default equation for higher values of q will be the same as the $q = 1$ equation.

For interfering signals within the receiver 80-dB bandwidth,

$$f_o - \frac{W}{2} \leq f_{INT} \leq f_o + \frac{W}{2}$$

$$P_{SR} \text{ (dBm)} = P_R \text{ (dBm)} + (160/W)\,[f - f_o].$$

For interference signals outside the receiver 80-dB bandwidth but within the overall tuning range of the receiver,

$$f_L \leq f_{INT} \leq f_o - W/2 \qquad \text{or} \qquad f_o + W/2 \leq f_{INT} \leq f_H$$

$$P_{SR} \text{ (dBm)} = P_R \text{ (dBm)} + 60 \text{ dB}.$$

For interfering signals outside the tuning range of the receiver,

$$f_{INT} < f_L \qquad \text{or} \qquad f_{INT} > f_H$$

$$P_{SR} \text{ (dBm)} = 0 \text{ dBm,}$$

where f_o is the receiver tuning frequency, W is the receiver 80-dB bandwidth, and P_R is the receiver sensitivity.

Chapter 5 | Propagation Effects on Interference

5.0 Introduction

As a result of the congestion of the frequency spectrum and the geostationary orbit and the related widespread use of frequency sharing, consideration of interference has assumed an important role in earth-station siting and other aspects of telecommunication system design. Interference may arise between terrestrial systems, between terrestrial and space systems, and between space systems. Attention is given here to interference involving space systems, whether between space systems or between space and terrestrial systems. Space-system earth stations, which commonly transmit high power and have sensitive receivers, may cause interference to terrestrial systems when transmitting and may be interfered with by terrestrial systems when receiving. In addition, one earth station may interfere with another. Also, earth stations may receive interfering, unwanted transmissions, as well as wanted signals, from satellites. Likewise, satellites may receive interfering transmissions from other than the intended earth station, and terrestrial systems may receive interference from space stations. In Section 5.1, some basic considerations are presented concerning the signal-to-interference ratio for a single wanted transmission and a single interfering transmission arriving over a direct path. The material presented in this chapter is based on the valuable research outlined in refs. [82–91].

In considering the problem of interference to or from an earth station, analysis may be separated into two stages. In the first, a coordination area surrounding the earth station is determined. This area, based on calculating coordination distances in all directions from the earth station, is defined such that terrestrial stations outside the area should experience or cause only a negligible amount of interference. To determine coordination distances, information on transmitter power, antenna gains, and permissible interference levels is needed. For the earth station, the aim toward the physical horizon on the azimuth considered is used. When considering interference due to scatter from rain, it is assumed that the beams of the two antennas intersect in a region where rain is falling. The coordination procedure is thus based on unfavorable assumptions with respect to mutual interference.

After the coordination area has been established, potential interference between the earth station and terrestrial stations within the coordination area can be

142

analyzed in more detail. In this stage of analysis, the actual antenna gains of the terrestrial stations in the directions toward the earth station will be used. Also, it is determined whether the beams of the earth station and terrestrial stations truly do intersect, in considering scatter from rain. Terrestrial stations within the coordination area may or may not be subject to or cause significant interference, depending on the factors taken into account in the second stage of analysis.

Two propagation modes are considered for determining coordination area. One involves propagation over near-great-circle paths, and one involves scatter from rain. Coordination distances d_1 and d_2 are determined for the modes and the larger of the two values is used as the final coordination distance.

From the propagation viewpoint, interference between terrestrial systems and earth stations is concerned very much with transhorizon propagation. In the late 1950s and early 1960s, transhorizon propagation became of considerable interest as a means of communication over long distances. The rather weak but consistent troposcatter signals were and are utilized for this purpose. The stronger but sporadic signals due to ducting and rain scatter do not occur for the high percentages of time needed for reliable communication, and much of the interest in transhorizon propagation at present is related to interference. Ducting and rain scatter contribute to the higher levels of interfering signals that occur for small percentages of time, and they are *highly* important in interference analysis [83]. The occurrence of ducting is vividly displayed on PPI radar screens showing ground clutter echoes.

5.1 Signal-to-Interference Ratio

The signal-to-noise ratio S/N of a telecommunication link is given by

$$(S/N)_{\text{dB}} = (\text{EIRP})_{\text{dBW}} - (L_{\text{FS}})_{\text{dB}} - L_{\text{dB}} + (G_R/T_{\text{sys}})_{\text{dB}} - k_{\text{dBW}} - B_{\text{dB}}. \qquad (5.1)$$

To consider this ratio, first separate EIRP into P_T and G_T, where EIRP stands for effective isotropic radiated power, P_T represents the transmitted power, and G_T represents transmitting antenna gain. Also, the loss factor L_{dB} can be separated into $A(p, \theta)$, attenuation in decibels expressed as a function of percentage of occurrence and elevation angle θ, and the factor $-20 \log \delta$ representing polarization mismatch [84]. As δ varies from 0 to 1, $-20 \log \delta$ is a positive quantity. Separating EIRP and L as indicated, S_{dBW} by itself becomes

$$S_{\text{dBW}} = (P_T)_{\text{dBW}} + (G_T)_{\text{dB}} + (G_R)_{\text{dB}} - (L_{\text{FS}})_{\text{dB}} - A(p, \theta) + 20 \log \delta. \qquad (5.2)$$

For I_{dBW}, the interfering power arriving over a direct path, a similar expression applies, namely

$$I_{dBW} = (P_{Ti})_{dBW} + (G_{Ti})_{dB} + (G_{Ri})_{dB} - (L_{FS})_{dB} - A_i(p,\theta) + 20 \log \delta_i, \qquad (5.3)$$

where the subscript i refers to the interfering signal. The quantity G_{Ti} represents the gain of the antenna of the interfering transmitter in the direction of the affected receiving system. A similar interpretation applies to the other terms. On the basis of Eqs. (5.2) and (5.3), the S/I ratio may be expressed as

$$(S/I)_{dB} = (P_T)_{dBW} - (F_{Ti})_{dBW} + (G_T)_{dB} - (G_{Ti})_{dB} + (C_R)_{dB}(G_{Ri})_{dB} \\ + 20 \log (d_i/d) + A_i(p,\theta) - A(p,\theta) + 20 \log (\delta/\delta_i). \qquad (5.4)$$

The term $20 \log (d_i/d)$ arises from the L_{FS} free-space basic transmission loss terms which have the form $(4\pi d/\lambda)^2$, where d is distance. In Eq. (5.4), d is the length of the path of the wanted signal and d_i is the length of the path of the interfering signal.

For analyzing transmissions from space to Earth or vice versa, the polarization mismatch factor δ equals $\cos \theta$, where θ is a polarization mismatch angle to which there may be three contributions such that

$$\theta = \theta_0 + \theta_i + \theta_r. \qquad (5.5)$$

The angle θ_0 arises from geometrical considerations and can be determined from

$$\theta_0 = \delta B - \alpha \delta A. \qquad (5.6)$$

δB is the difference in back azimuths between the service path (to the intended earth station) and the interfering path (to the earth station being interfered with). The back azimuth is the angle to the earth station measured from the north–south meridian of the subsatellite point. The factor δA represents the difference in azimuths of the two earth stations, azimuth in this case being measured at the earth station as the angle from geographic north, to the great-circle path from the earth station to the subsatellite point (Figure 5.1). The quantity A depends on the great-circle distance Z between the earth stations. On this topic, we follow the treatment by Dougherty [84] and reproduce two of his illustrations showing θ_0 as a function of B (Figure 5.2) and B and Z as a function of earth station latitude and longitude with respect to the subsatellite point (Figure 5.3).

The angle θ_i represents the Faraday rotation of a linearly polarized wave that may take place in propagation through the ionosphere. The concept of Faraday

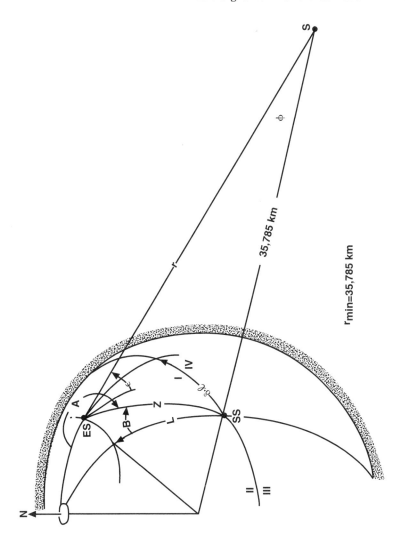

Figure 5.1 Synchronous satellite geometry.

rotation is not applicable to circularly polarized waves. The relation for θ_i used by Dougherty [84] is

$$\theta_i = 108^0/f^2,$$

with f the frequency in GHz. This value of θ_i corresponds to the maximum one-way effect of the ionosphere for an elevation angle of 30°. The angle θ_r represents

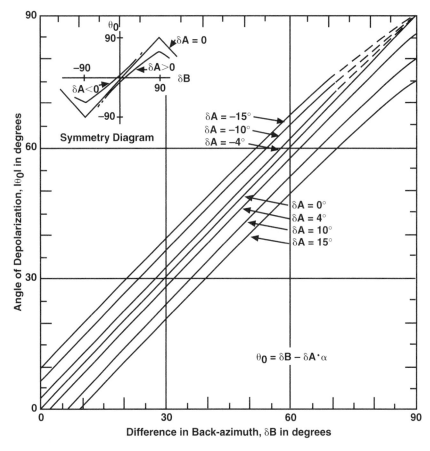

Figure 5.2   The depolarization angle for linear polarization for a potential interference
situation.

the possible rotation of the electric field intensity due to depolarization caused
by precipitation or other effects. By definition, cross polarization discrimination
(XPD) is given by

$$\text{XPD} = 20 \log (E_{11} / E_{12})$$

where E_{11} is the amplitude of the copolarized signal (having the original polarization and after taking account of any attenuation along the path) and E_{12} is the

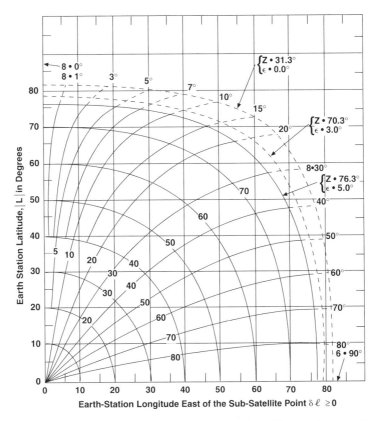

Figure 5.3 Great-circle arc (Z) and back-azimuth (B from SS to ES) as a function of the earth-station (ES) latitude L and degrees of longitude ($\delta\ell$) east of the subsatellite point (SS).

amplitude of the orthogonally polarized signal produced by depolarization. The angle θ_r is $\tan^{-1} E_{11}/E_{12}$.

For determining the values of $A(p,\theta)$ and δ in (5.2) and (5.3), one evaluates the service path under unfavorable conditions, using the loss occurring for a small percentage of the time, corresponding to $p = 0.01\%$ for example. The interference path, however, is evaluated with the minor losses occurring, say, 50% of the time. This practice takes into account such possibilities as the wanted signal propagating through an intense rain cell while the unwanted signal follows a path which misses the rain cell and encounters negligible attenuation.

5.2   Coordination Area Based on Great-Circle Propagation

5.2.1 BASIC CONCEPTS

For determining coordination area, attenuation needs to be estimated for the two modes of propagation of interfering signals [85–87]. Propagation mode 1, referring to propagation over a direct near-great-circle path, occurs essentially all of the time to some degree. The second propagation mode is primarily via scatter from rain and may occur infrequently.

In system planning, one is generally required to estimate the relatively intense interference level exceeded for some small percentage p of the time (e.g., $p = 0.01\%$) and also perhaps the interference level exceeded for about 20% ($p = 20\%$) of the time. Corresponding to high interference levels are values of basic transmission loss L_b (Figure 5.4). Note that in considering attenuation due to rain, the concern is for the small percentages of time for which the highest interfering signal intensities occur.

The total loss factor L_t relating the transmitted interfering power P_{ti} and the received interfering power P_{Ri} is defined by

$$L_t = \frac{P_{Ti}}{P_{Ri}}. \tag{5.7}$$

An expression for the basic transmission loss, L_b, referred to earlier can be obtained from $P_{Ri} = P_{Ti}\, G_{Ti}\, G_{Ri}/L_{FS}\, L$, identifying $L_{FS}\, L$ as L_b,

$$L_b = L_{FS}\, L = \frac{P_{Ti}\, G_{Ti}\, G_{Ri}}{P_{Ri}}, \tag{5.8}$$

where L_{FS} is the free-space basic transmission loss and L represents other system losses. In decibel values referring to p percent of the time, Eq. (5.7) becomes

$$[L_t(P)]_{dB} = (P_{Ti})_{dBW} - [P_{Ri}(p)]_{dBW} \tag{5.9}$$

and Eq. (5.8) becomes

$$[L_b(p)] = (P_{Ti})_{dBW} + (G_T)_{dB} + (G_R)_{dB} - [P_{Ri}(P)]_{dBW}. \tag{5.10}$$

In Eqs. (5.9) and (5.10), $P_{Ri}(p)$ is the maximum permissible interfering power level to be exceeded for no more than p percent of the time. The gains G_T and G_R are the gains of the transmitting and receiving antennas. For determining coordination distance, the horizon gain at the azimuth in question is used for the earth-satellite station and the maximum gain is used for the terrestrial station.

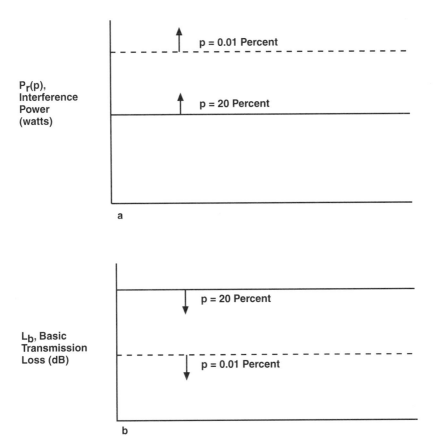

Figure 5.4 Correspondence between interference level and basic transmission loss. The interfering signal power will be above a certain level 0.01% of the time, as suggested by the arrow extending upwards from the dotted line in (a). The high interfering levels above the dotted line of (a) correspond to the low values of basic transmission loss below the dotted line of (b). For 20% of the time the interfering level will be above the solid line of (a), and the corresponding values of basic transmission loss will be below the solid line of (b).

From Eq. (5.8), it can be seen that if $G_T = G_R = 1$, then $L_b = P_{Ti}/G_{Ri}$. For this reason, L_b is said to be the loss that would occur between isotropic antennas.

The basic transmission loss L_b is seen to be the product of L_{FS} and L. For a line-of-sight path and for frequencies below 10 GHz, L_b is roughly but not necessarily exactly equal to L_{FS}. In any case, L_{FS} makes a major contribution to L_b. The free-space transmission loss is defined by

$$L_{FS} = (4 \pi d / \lambda)^2, \tag{5.11}$$

where d is the distance from transmitting to receiving locations and λ is wavelength. At higher frequencies, the dissipative attenuation associated with water vapor and oxygen may make significant contributions to L_b. Attenuation of the interfering signal due to rain is not included in L_b, for the low values of p normally considered in applying Eq. (5.11) as $L_b(p)$ then represent the low values of basic transmission loss that can be tolerated for only small percentages of time. When considering interfering signals, high values of L_b can be readily tolerated. It is the low values of L_b that are of concern. In terms of decibel values, Eq. (5.11) can be written as

$$(L_{FS})_{dB} = 20 \log(4\pi) + 20 \log d - 20 \log \lambda, \tag{5.12}$$

where d and λ are in meters. Commonly, however, L_{FS} is expressed in terms of frequency f rather than wavelength λ. By replacing λ with c/f, where $c = 2.9979 \times 10^8$ m/sec, one obtains

$$(L_{FS})_{dB} = -147.55 + 20 \log f + 20 \log d. \tag{5.13}$$

If f is expressed in gigahertz rather than hertz, a factor of 180 dB must be added to the right-hand side of Eq. (5.13), and if d is expressed in kilometers rather than meters, an additional factor of 60 dB must also be included, with the result that

$$(L_{FS})_{dB} = 92.45 + 20 \log f_{GHz} + 20 \log d_{km}. \tag{5.14}$$

5.2.2   *LINE-OF-SIGHT PATHS*

Although L_b may equal L_{FS} approximately for frequencies below 10 GHz for a certain range of values of p, in the absence of horizon or obstacle effects, the actual received interfering signal on even a clear line-of-sight path fluctuates because of the effects of atmospheric multipath propagation, scintillation, and defocusing and may be greater or less than L_{FS}. Thus, although L of Eq. (5.8) has been referred to as a loss factor, it must be able to assume values either greater or less than unity if it is to be applicable to the situation considered here. The variation of the received level P_{Ri} with time provides the basis for specifying P_{Ri} as a function of p. For line-of-sight paths, L can be expressed as $A_o + A_d - G_p$, and L_b is given by

$$(L_b)_{dB} = (L_{FS})_{dB} + A_o0 + A_d - G_p, \qquad (5.15)$$

where A is attenuation in decibels due to oxygen and water vapor. (That due to water vapor can be neglected below about 15 GHz.) The coefficient A_d represents attenuation due to defocusing in decibels, and G_p is an empirical factor in decibels given by Table 5.1 for paths of 56 km or greater. For distances shorter than 50 km, the values of G_p are proportionally reduced. To estimate the signal exceeded for percentages of the time between 1 and 20, CCIR Report 569 [85] recommends adding 1.5 dB to the value of L_{FS} (thereby increasing L_b by 1.5 dB with respect to what it would be otherwise). The coefficient G_p can be taken to be zero for $p = 20\%$ and greater.

Attenuation due to defocusing results when the variation of refractivity with height dN/dh itself varies with height, so that rays at different heights experience different amounts of bending. Rays representing energy propagation, rays which were originally essentially parallel, for example, may then become more widely separated than otherwise and signal intensity is consequently reduced. It develops that the variation of dN/dh with height h is proportional to ΔN, the decrease in refractivity N in the first kilometer above the surface. Figure 5.5 shows attenuation due to defocusing as a function of ΔN and elevation angle θ [88]. A given path may be a clear line-of-sight path for certain values of dN/dh but may have part of the first Fresnel zone obstructed for other values of dN/dh.

5.2.3 TRANSHORIZON PATHS

Major attention in the analysis of interference between terrestrial systems and earth stations of space systems is directed to transhorizon propagation. The term transhorizon path refers to a path extending beyond the normal radio horizon for which diffraction is a relevant propagation mechanism, as distinguished from a clear line-of-sight path at one extreme and a strictly troposcatter path at the

Table 5.1 G_p **of Eq. (5.15) vs Percent of Time p Exceeded**

p (percent)	0.001	0.01	0.1	1
G_p (dB)	8.5	7.0	6.0	4.5

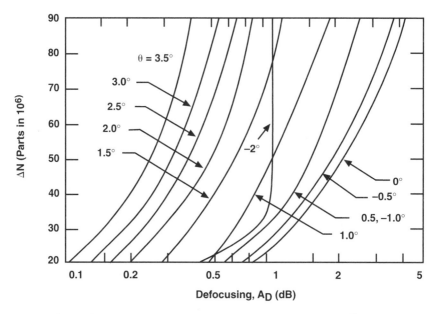

Figure 5.5 Defocusing on near-horizontal paths as a function of ΔN.

opposite extreme. For transhorizon paths, a diffraction loss term A_s (dB) must be added to the free-space loss I_{FS}. In addition, account must be taken of ducting and super-refraction, which can be expected to occur for some percentage of the time. The path loss for transmitting and receiving terminals which are both immersed in a duct is

$$(L_b)_{dB} = 92.45 + 20 \log f_{GHz} + 10 \log d_{km} + A_c$$
$$+ (\gamma_d + \gamma_0 + \gamma_w) d \text{ (km)} + A_s. \tag{5.16}$$

This equation includes terms like those of Eq. (5.14) for L_{FS}, except that 10 log d appears instead of 20 log d. The basis for using 10 log d is that a wave in a duct is constrained in the vertical direction and spreads out only horizontally, whereas in free space a wave spreads in both directions. Because L_b for a duct includes 10 log d rather than 20 log d, L_b tends to be significantly less than L_{FS}. The quantity A_c represents a coupling loss that takes account of the fact that not all the rays leaving the transmitting antenna are trapped within the duct. The γ's are attenuation constants, γ_d being a duct attenuation constant reported to have a theoretical minimum value of 0.03 dB/km [89]. The constants γ_0 and γ_w

represent attenuation due to oxygen and water vapor, respectively. The quantity A_s takes account of loss caused by obstacles along the path. CCIR Reports 382-5 [87] and 724-2 [86], however, use, for L_b for ducting,

$$(L_b)_{dB} = 120 + 20 \log f_{GHz} + \gamma_d \text{ (km)} + A_h. \tag{5.17}$$

The term γ includes the γ's of Eq. (5.16), and A_h is a modified form of A_s of Eq. (5.16). Equation (5.16) has the advantage of being closely related to the physical phenomena involved, but it has the comparison disadvantage of having a term involving the logarithm of distance and a term that is linear with distance. One needs to solve for d, the coordination distance for great-circle propagation, and for this purpose Eq. (5.17) has the advantage of having only a term that is linear with distance. The basis for the conversion from Eq. (5.16) to (5.17) is that the term $10 \log d$ can be approximated by

$$10 \log d \text{ (km)} = 20 + 0.01 \, d \text{ (km)}, \quad 100 \text{ km} < d < 2000 \text{ km}. \tag{5.18}$$

Also, the coupling loss A_c of Eq. (5.16) has been assigned the value of 7.5 dB, whereas in CCIR Report 569-3 [85] this loss is given by a table showing it as varying from 6 to 11 dB over water and coastal areas and 9 to 14 dB over inland areas. The value of 120 is obtained by setting 92.45 equal to 92.5 and noting that $92.5 + 20 + 7.5 = 120$. The coefficient 0.01 of Eq. (5.18) is included as part of the γ of Eq. (5.17), and γ is then given by

$$\gamma = 0.01 + \gamma_d + \gamma_0 + \gamma_w. \tag{5.19}$$

The quantity A_s of Eq. (5.16), expressed in decibels, has the form

$$A_s = 20 \log [1 + 6.3 \, \theta \, (f \, d_h)^{1/2}] + 0.46 \, \theta \, (f \, C_r)^{1/3} \tag{5.20}$$

where f is frequency in gigahertz, d_h is distance to the horizon in kilometers, θ is elevation angle in degrees above the horizon, and C_r is the radius of curvature of the horizon. If d_h is set equal to 0.5 km and C_r is taken to be 10 m, one obtains the horizon angle correction A_h of Eq. (5.17), namely,

$$A_h = 20 \log (1 + 4.5 f^{1/2} \, \theta) + f^{1/3} \, \theta \tag{5.21}$$

Figure 5.6 shows A_h as a function of elevation angle and frequency. The factor γ_d is given by [86]

$$\gamma_d = [c_1 + c_2 \log (f + c_3)] \, p^{C_4} \text{ (dB/km)}, \tag{5.22}$$

where the c's have different values for four different zones and are given in Table 5.2. The frequency f is in gigahertz, and p is percentage of time.

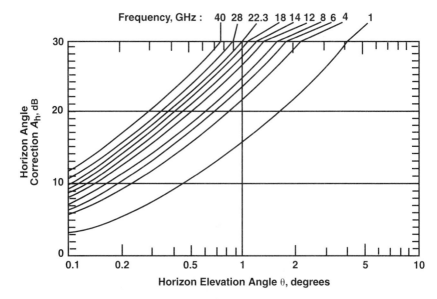

Figure 5.6 The horizon angle correction, A_h [Eq. (5.21)].

Table 5.2 **Values of Constants for Determination of γ_d**

	c_1	c_2	c_3	c_4
Zone A1	0.109	0.100	−0.10	0.16
Zone A2	0.146	0.148	−0.15	0.12
Zone B	0.050	0.096	0.25	0.19
Zone C	0.040	0.078	0.25	0.16

The zones referred to in the table are as follows:

Zone A1: Coastal land and shore areas, adjacent to zones B or C to an elevation of 100 m relative to mean water level, but limited to a maximum distance of 50 km from the nearest zone B or C area.

Zone A2: All land, other than coastal land and shore areas.

Zone B: ''Cold'' seas, oceans, and other substantial bodies of water, encompassing a circle 100 km in diameter at latitudes greater than 23.5° N or S, but excluding all of the Black Sea, Caribbean Sea, Gulf of Mexico, Mediterranean Sea, Red Sea, and the sea from the Shatt-al-Arab to and including the Gulf of Oman.

Zone C: "Warm" seas, oceans, and other substantial bodies of water, encompassing a circle 100 km in diameter, and including in their entirety the bodies of water mentioned as being excluded from zone B.

The constant γ_0 for oxygen is given in CCIR Report 724-2 [86] in dB/km for $f < 40$ GHz by

$$\gamma_0 = \left[0.00719 + \frac{6.09}{f^2 + 0.227} + \frac{4.81}{(f - 57)^2 + 1.50} \right] \frac{f^2}{1000}. \quad (5.23)$$

Attenuation due to water vapor can be neglected for frequencies less than 15 GHz, and the expression for γ_W is therefore not given here.

CCIR Report 724-2 [86] includes plots for a graphical solution for coordination distance for ducting, or great-circle propagation. We do not include these illustrations here, but Eq. (5.17) can be solved algebraically for the distance d by making use of the accompanying information about the parameters appearing in it.

Troposcatter signals, resulting predominantly from inhomogeneous scattering by random fluctuations of the index of refraction of the atmosphere, are normally weaker than the interfering signals due to ducting and super-refraction. However, the tropospheric scatter signals may be dominant for percentages of time between about 1 and 50% and for percentages less than 1% when high site shielding (A_h values of 30 dB and greater) is encountered.

5.3 Coordination Area for Scattering by Rain

For considering interference due to scatter from rain, one can start with a slightly modified version of an equation which refers to bistatic scatter from rain. Inverting this relation to obtain a total loss factor L_t using G_T, G_{ES}, R_T, and R_{ES} to refer to the gains of the terrestrial and earth station antennas and their distances from the region of rain scatter, and replacing W_T and W_R by P_T and P_R, results in

$$L_t = \frac{P_{Ti}}{R_{Ri}} = \frac{(4\pi)^3 R_T^2 R_{ES}^2 L}{G_T G_{ES} \eta V \lambda^2}. \quad (5.24)$$

In this expression, L is a loss factor (greater than unity), V is the common scattering volume, and η is the radar cross-section per unit volume. For Rayleigh scattering η has the form

$$\eta = \frac{\pi^5}{\lambda^4} \left| \frac{K_c - 1}{K_c + 2} \right|^2 Z \quad (\text{m}^2/\text{m}^3), \quad (5.25)$$

where K_c is the complex dielectric constant of water and is a function of frequency and temperature. When expressed in mm^6/m^3, the quantity Z is related to rainfall rate R in mm/h for a Laws and Parsons distribution of drop sizes by the empirical expression

$$Z = 400 \ R^{1.4}. \tag{5.26}$$

Physically, Z represents Σd^6, where d is the drop diameter and the summation is carried out for all of the drops in a unit volume. For frequencies higher than 10 GHz for which Rayleigh scattering does not apply, an effective of modified value of Z, designated as Z_e, is used for coordination distance calculations.

Usually the earth-station antenna has a smaller beamwidth than the terrestrial antenna. Assuming that such is the case and noting that the scattering volume is defined by the antenna with the smallest beamwidth, V is given approximately by

$$V = (\pi/4) \ \theta^2 \ R^2_{ES} \ D, \tag{5.27}$$

where θ is the beamwidth of the earth-station antenna, R_{ES} is the distance from the earth station to the common scattering volume V, and D is the extent of the common scattering volume along the path of the earth-station antenna beam. Assuming a circular aperture antenna for which the beamwidth θ is given approximately by λ/d, where d is diameter, and making use of the relation between effective antenna area A and gain G, namely $G = 4\pi A/\lambda^2$, it develops that $\theta^2 = \pi^2/G$ and

$$V = \pi^3 \ R^2_{ES} \ D / (4 \ G_{ES}). \tag{5.28}$$

Substituting for η and V in Eq. (5.24) and recognizing that $\eta|(K_c - 1)/(K_c + 2)|$ has a value of about 0.93 [90],

$$L_t = \frac{256 \ R^2_T \ L\lambda^4}{G_T \ D\lambda^2 \ \pi^5 \ (0.43)Z} \tag{5.29a}$$

Combining the numerical factors of Eq. (5.29a) and replacing λ by c/f results in

$$L_t = \frac{0.9R^2_T \ c^2L}{f^2G_T \ DZ}. \tag{5.29b}$$

Note that R_{ES} and G_{ES} have dropped out of the expression for L_t, but that R_T and G_T remain. Taking logarithms results in

$$
\begin{aligned}
(L_t)_{dB} &= 0.46 + 20 \log R_T + 169.54 + 10 \log L \\
&\quad - 20 \log f - 10 \log G_T - 10 \log D - 10 \log Z \\
(L_t)_{dB} &= 199 + 20 \log (R_T)_{km} + 10 \log L \\
&\quad - 20 \log f_{GHz} - 10 \log D_{km} - 10 \log Z_{mm^6/m^3} - 10 \log G_T,
\end{aligned}
\tag{5.30}
$$

where $+60$ is introduced when replacing R_T in meters by R_T in kilometers, and -30 is introduced when replacing D in meters by D in kilometers. Changing from f in hertz to f in gigahertz and from Z in m^6/m^3 to Z in mm^6/m^3 introduces two 180-dB factors of opposite sign, which cancel out. The relation of Eq. (5.30) can be modified to express D and Z in terms of rain rate R. The distance D is taken to be given by

$$D = 3.5 \ R^{-0.08}, \tag{5.31}$$

based on modeling of rain cells and assuming an elevation angle of $20°$ as a conservative assumption. For Z, assuming a Laws and Parsons distribution of drop sizes,

$$Z = 400 \ R^{1.4}$$

Taking $10 \log D$, one obtains $5 - 0.8 \log R$, and taking $10 \log Z$ gives $26 + 14 \log R$. Subtracting $26 + 5$ from 199 leaves 168, and combining the $\log R$ terms results in $-13.2 \log R$. The resulting equation, derived from Eq. (5.30), after also specifying the contributions to L, is

$$\begin{aligned}(L_t)_{dB} = \ &168 + 20 \log (R_T)_{km} - 20 \log f_{GHz} - 13.2 \ R \\ &- 10 \log G_T - 10 \log C + \gamma_o r_o + \Gamma.\end{aligned} \tag{5.32}$$

The quantity C accounts for attenuation in the common scattering volume. The expression for C given in CCIR Report 724-2 [86] is

$$C = [2.17/(\gamma_r D)] \ (1 - 10^{-\gamma_r D/5}), \tag{5.33}$$

where γ_r is the attenuation constant for rain for vertical polarization. D, the path through rain, is defined by Eq. (5.31), and $\gamma_o r_o$ is attenuation due to oxygen. The distance r_o is an effective distance equal to $0.7 \ R_T + 32$ km for $R_T < 340$ km and otherwise 270 km. The quantity Γ represents attenuation due to rain outside the common scattering volume. It is given by a rather complicated expression in CCIR Report 724-1:

$$\Gamma = 631 \ kR^{\alpha - 0.5} \ 10^{-(R+1)0.19}. \tag{5.34}$$

Equation (5.34) can be solved for R_T, the distance from the common scattering volume to the terrestrial station. The distance R_T, however, is not the rain-scatter coordination distance d_2, as R_T is not measured from the earth station. The center of the circle representing the locus of R_T (scatter is assumed to occur equally in all directions from the common scattering volume) is displaced from the earth station by Δd, which is a function of elevation angle θ, where

$$\tan \theta = \frac{h}{\Delta d} = \frac{(R_T - 40)^2}{17,000 \Delta d} \tag{5.35}$$

and

$$\Delta d = \frac{(R_T - 40)^2 \cot \theta}{17,000} \tag{5.36}$$

The basis for this relation is shown in Figure 5.7. The grazing ray from the terrestrial transmitter is assumed to graze the horizon at a distance of 40 km, and a k factor of 4/3 is assumed. The expression in CCIR Report 382-5 [87] that corresponds to Eq. (5.37) has the same form, except that a gain G_T of 42 dB is assumed, and $168 - 42 = 126$, so that for $f \leq 10$ GHz,

$$\begin{aligned}(L_t)_{dB} = {}&126 + 20 \log (R_T)_{km} - 20 \log f_{GHz} - 13.2 \log R \\ &- 10 \log C + \gamma_o r_o + 10 \log B,\end{aligned} \tag{5.37}$$

where $10 \log B$ takes the place of Γ but has the form of Γ [90].

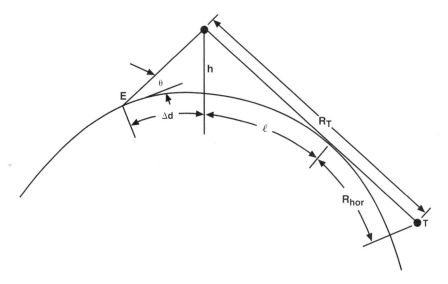

Figure 5.7 Rain scatter involving a transhorizon path from a terrestrial station T. A grazing ray at the horizon will reach a height of $l^2/2kr_0 = (R_T - R_{hor})^2/2kr_0$ at the distance $R_T - R_{hor}$ from the ray, which is tangential to the earth's spherical surface. The elevation angle θ corresponding to this height h, as seen from the earth station E, is $\tan^{-1} h/\Delta d$.

Another variation of the equation for interference caused by scatter from rain is

$$[L_2 (0.01)]_{dB} = 131 - 20 \log (R_T)_{km} - 20 \log f_{GHz} - 10 \log C$$
$$+ \gamma_o r_o - 14 \log R + (R_t - 40)^2/17{,}000 - 10 \log D_{km}.$$
$$(5.38)$$

The loss in this case is for a percentage of occurrence of 0.01. The 10 log D_{km} term is retained as such and the "5" referred to following Eq. (5.26) does not appear, so the numerical coefficient of Eq. (5.38) is 131 rather than 126. Also, the quantity Z is assumed to decrease at a rate of 1 dB/km, and this decrease is accounted for by subtracting h of Eq. (5.36) from Z [$h = (R_T - 40)^2/17{,}000$]. As it is $-10 \log Z$ that occurs in the original equation, Eq. (5.38) includes $+h$.

5.4 Interference between Space and Surface Stations

Interference between a space station and one on the earth's surface may take place, for example, when an earth station receives unwanted transmissions from an interfering satellite as well as wanted transmissions from the satellite that serves the earth station. The analysis of Section 5.2, presented as an introduction to the analytical aspects of interference, applies directly to this case, and some additional considerations follow. Because the spacings of satellites in the geostationary orbit may be as close as 2°, limitations on the uplink and dowlink antenna gains off-axis have been prescribed by the FCC. Uplink antenna gain is limited to $32 - 25 \log \theta$, where θ is the off-axis angle in degrees, for values of θ of 1° and greater. For downlinks, the corresponding expression is $29 - 25 \log \theta$. A different approach to combat interference, however, is to use the spread-spectrum technique. Small earth-station antennas can then be employed, and discrimination against unwanted signals can be obtained by using code-division multiple access.

Scatter from rain, which was not considered in Section 5.1 but may also cause interference, can be analyzed by a modification of the approach of Section 5.3 with R_T and G_T now taken to refer to the interfering satellite transmitter rather than to a terrestrial transmitter.

Solar power satellites, which would intercept solar energy and transmit energy to the earth's surface as microwave radiation at a frequency of 2450 MHz according to preliminary plans, present a potential interference problem for communication satellite systems. According to one analysis [88] based on likely harmonic content, the interfering signal scattered from rain, even at the fourth harmonic, would be comparable with the signal level received in the fixed satellite service.

In the absence of precipitation, the signal on a line-of-sight path from a satellite will be attenuated by the atmospheric gases and perhaps by defocusing, but may experience a gain due to multipath and scintillation effects, for a small fraction of the time, as mentioned in Section 5.3.2. The gain due to multipath effects and scintillation may be assumed to be zero for elevation angles above 5° and percentages of time greater than 1% [88].

5.5   Procedures for Interference Analysis

5.5.1   INTRODUCTION

Previous sections of this chapter have outlined the theoretical basis for interference analysis, with emphasis on basic concepts. In this section, practical considerations, including procedures for determining coordination distance, are summarized.

5.5.2   OFF-AXIS ANTENNA GAIN

For calculating the predicted intensity of a terrestrial interfering signal at an earth station or of an interfering signal from an earth station at a terrestrial station, it is necessary to know the gain of the earth station antenna at the horizon, at the azimuth of the terrestrial station (or for determining coordination distance at all azimuthal angles). To determine the gain, one must first find the angle of the horizon from the axis of the main antenna beam at the azimuth of interest. For the case that the horizon is at zero elevation angle, the horizon angle θ, measured from the axis of the antenna beam, is found by applying the law of cosines for sides of a spherical triangle, namely,

$$\cos \theta = \cos \theta_s \cos (\alpha - \alpha_s),$$

where θ_s is the elevation angle of the satellite the earth station is servicing, α_s is the azimuth of the satellite, and α is the azimuthal angle of interest. If the horizon is at an elevation angle θ, the corresponding relation becomes

$$\cos \phi = \cos \theta \cos \theta_s \cos (\alpha - \alpha_s) + \sin \theta \sin \theta_s. \qquad (5.39)$$

Having determined ϕ, it remains to specify a value for the antenna gain at this angle. If the actual antenna gain is known as a function of ϕ, it should be used. If the gain is not known and the antenna diameter to wavelength ratio D/λ is 100 or greater, the following relation can be used for angles in degrees greater than that of the first sidelobe:

$$G = 32 - 25 \log \phi \text{ (dB)}. \qquad (5.40)$$

If the D/λ ratio is less than 100, the corresponding relation is

$$G = 52 - 10 \log (D/\lambda) - 25 \log \phi. \qquad (5.41)$$

The same sources give relations between the maximum gain G_{max} and D/λ:

$$20 \log D/\lambda = G_{max} - 7.7 \text{ (dB)}. \qquad (5.42)$$

More precisely and completely than stated earlier, the following set of relations are given for $D/\lambda \geq 100$:

$$
\begin{aligned}
G(\phi) &= G_{max} - 2.5 \times 10^{-3} (D\phi/\lambda) & \text{for } 0 < \phi < \phi_m & \qquad (5.43a) \\
G(\phi) &= G_1 & \text{for } \phi_m < \phi < \phi_r & \qquad (5.43b) \\
G(\phi) &= 32 - 25 \phi & \text{for } \phi_r < \phi < \phi_m & \qquad (5.43c) \\
G(\phi) &= -10 & \text{for } 48° < \phi < 180°, & \qquad (5.43d)
\end{aligned}
$$

where

$$
\begin{aligned}
\phi_m &= (20\lambda/D) (G_{max} - G_1)^{0.5} \text{ (degree)} \\
\phi_r &= 15.85 (D/\lambda)^{-0.6} & \qquad (5.44) \\
G_1 &= 2 + 15 \log D/\lambda.
\end{aligned}
$$

For $D/\lambda \leq 100$,

$$
\begin{aligned}
G(\phi) &= G_{max} - 2.5 \times 10^{-3} (D\phi/\lambda) & \text{for } 0 < \phi < \phi_m & \qquad (5.45a) \\
G(\phi) &= G_1 & \text{for } \phi_m < \phi < 100 \, \lambda/d & \qquad (5.45b) \\
G(\phi) &= 52 - 10 \log D/\lambda - 25 \log \phi & \text{for } 100\lambda/D < \phi < 48 & \qquad (5.45c) \\
G(\phi) &= 10 - 10 \log D/\lambda & \text{for } 48° < \phi < 180°. & \qquad (5.45d)
\end{aligned}
$$

For satellite antennas we have the relations

$$
\begin{aligned}
G(\phi) &= G_{max} - 3(\phi/\phi_0)^2 & \phi_0 < \phi < a\phi_0 & \qquad (5.46a) \\
G(\phi) &= G_{max} + L_s & a\phi_0 < \phi < 6.32 \, \phi_0 & \qquad (5.46b) \\
G(\phi) &= G_{max} + L_s + 20 - 25(\phi/\phi_0) & 6.32 \, \phi_0 < \phi < \phi_1 & \qquad (5.46c) \\
G(\phi) &= 0 & \phi_1 < \phi,
\end{aligned}
$$

where ϕ_0 is one-half the 3-dB beamwidth and ϕ_1 is the value of ϕ when $G_{max} = 0$. The parameter a has values of 2.58, 2.88, and 3.16 when L_s, the required near-in sidelobe level relative to the peak, has the values -20, -25, and -30 dB, respectively.

5.5.3 PROCEDURES FOR DETERMINING COORDINATION FOR GREAT-CIRCLE PROPAGATION

For determining coordination distances d_1 for great-circle propagation, it is necessary to first determine the basic transmission loss, L_b, as defined by Eq. (5.10), that can be tolerated for the percentage of time specified (commonly 0.01% and perhaps 20% as well). The allowable value of L_b is based primarily on factors other than propagation. The quantity $P_{Ri}(p)$ should be taken to be the maximum permissible interference level for p percent of the time.

If the interfering station is an earth station, the gain toward the physical horizon on the azimuth in question is to be used. If the interfering station is a terrestrial station, the maximum expected antenna gain is to be used. The quantity G_R refers to the gain of the station that is interfered with. If the station is an earth station, the gain toward the horizon on the azimuth in question is to be used. If the station experiencing interference is a terrestrial station, the maximum expected antenna gain is to used. For determining coordination distance for installation of an earth station, one can initially determine coordination distance in all directions without regard to locations of terrestrial stations. In a second stage of analysis after coordination distance has been determined, the locations and gains of the terrestrial stations toward the earth station can be utilized to determine if an interference problem truly exists.

Note that antenna gains were taken into account in determining the value of L_b of Eqs. (5.16) and (5.17) but do not appear explicitly in either of the two equations. The coordination distance found from these equations is designated as d_1. The reports cited include descriptions of procedures for use when great-circle paths cross more than one zone.

For zones B and C (Section 5.2.3), if coordination distances turn out to be greater than the values in Table 5.3, the values in the table should be used instead as the coordination distance.

Table 5.3 **Maximum Coordination Distance d_1**

	Percent of Time			
Zone	**0.001**	**0.01**	**0.1**	**1.0**
B	2000 km	1500 km	1200 km	1000 km
C	2000 km	1500 km	1200 km	1000 km

5.6   Permissible Level of the Interfering Emissions

Information on the permissible level of interfering emission that is included in ITU is reproduced next. The permissible level of the interfering emission (dBW) in the reference bandwidth, to be exceeded for no more than $p\%$ of the time at output of the receiving antenna of a station subject to interference, from each source of interference, is given by

$$P(p) = 10 \log(kT_eB) + J + M(p) - W, \tag{5.47}$$

where

$$M(p) + M(p_0/n) = M_0(p_0) \tag{5.48}$$

with

K: Boltzmann's constant (1.38×10^{-23} J/K)

T_e: thermal noise temperature of the receiving system (K) at the output of the receiving antenna

B: reference bandwidth (Hz) (bandwidth of the interfered-with system over which the power of the interfering emission can be averaged)

J: ratio (dB) of the permissible long term (20% of the time) interfering emission power to the thermal noise power of the receiving system, referred to the output terminals of the receiving antenna

p_0: percentage of the time during which the interferences from all sources may exceed the permissible value

n: number of expected entries of interference, assumed to be uncorrelated

p: percentage of the time during which the interference from one source may exceed the permissible value, since the entries of interference are not likely to occur simultaneously: $p = p_0/n$

$M_0(p_0)$: ratio (dB) between the permissible powers of the interfering emission, during $p_0\%$ and 20% of the time, respectively, for all entries of interference

$M(p)$: ratio (dB) between the permissible powers of the interfering emissions during $p\%$ of the time for one entry of the interference, and during 20% of the time for all entries of interference

W: equivalence factor (dB) relating interference from interference emissions to that caused by the introduction of additional thermal noise of equal power in the reference bandwidth; it is positive when the interfering emissions would cause more degradation than thermal noise

The noise temperature in kelvins of the receiving system referred to the output terminals of the receiving antenna may be determined from

$$T_e = T_0 + (L - 1)290 + LT_r, \tag{5.49}$$

where T_e is the noise temperature (K) contributed by the receiving antenna, and L is the numerical loss in the transmission line (e.g., waveguide) between antenna and receiver front end. T_r is the noise temperature (K) of the receiver front end, including successive stages, referred to the front end input.

The factor J (dB) is defined as the ratio of total permissible long-term (20% of the time) power of interfering emissions in the system, to the long-term thermal radio frequency noise power in a single receiver. In the computation of this factor, the interference emission is considered to have a flat power spectral density, its actual spectral shape being taken into account by the factor W. In digital systems interference is measured and prescribed in terms of the bit error rate or its permissible increase. Although the bit error rate increase is additive in a reference circuit comprising tandem links, the radio frequency power of interfering emissions giving rise to such bit error rate increase is not additive, because bit error rate is not a linear function of the level of the radio frequency power of interfering emissions. Thus, it may be necessary to protect each receiver individually. For digital radio relay systems operating above 10 GHz and for all digital satellite systems, the long-term interference power may be of the same order of magnitude as the long-term thermal noise; hence, $J = 0$ dB. For digital radio-relay systems operating below 10 GHz, long-term interference power should not decrease the receiver fade margin by more than 1 dB. Thus, the long-term interference power should be about 6 dB below the thermal noise power, and hence, $J = -6$ dB.

$M_0(p_0)$ (dB) is the interference margin between the short-term (p_0%) and long-term (20%) allowable power of an interfering emission. In the case of digital systems, system performance at frequencies above 10 GHz can, in most areas of the world, usefully be defined as the percentage of the time p_0 for which the wanted signal is allowed to drop below its operating threshold, defined by a given bite error rate. During nonfaded operation of the system, the desired signal will exceed its threshold level by some margin M1 which depends on the rain climate in which the stations operate. The greater this margin, the greater the enhancement of the interfering emissions which would degrade the system to threshold performance. As a first approximation it may be assumed that, for small percentages of the time, the level of interfering emissions may be allowed to equal the thermal noise which exists at the demodulator input-faded conditions. For digital radio-relay systems operating below 10 GHz it is assumed that the short-term power of an interfering emission can be allowed to exceed the long-term power of the interfering emission by an amount equal to the fade margin of the system minus J.

The factor W (dB) is the ratio of radio frequency thermal noise power to the power of an interfering emission in the reference bandwidth when both produce the same interference after demodulation. For FM signals, it is defined as follows:

$$W = 10 \log |T_1 / P_1| |P_2 / T_2|, \tag{5.50}$$

where T_1 is the thermal noise power at the output of the receiving antenna in the reference bandwidth, P_1 is the power of the interfering emission at the radio frequency in the reference bandwidth at the output of the receiving antenna, P_2 is the interference power in the receiving system after demodulation, and T_2 is the thermal noise power in the receiving system after demodulation.

The factor W depends on the characteristics of the wanted and the interfering signals. To avoid the need for considering a wide range of characteristics upper limit values were determined for the factor W. When the wanted signal uses frequency modulation with r.m.s modulation indices which are greater than unity, W is not higher than 4 dB. In such cases, a conservative figure of 4 dB will be used for the factor W regardless of the characteristics of the interfering signals.

5.7   Link Communications Systems Design

The role of propagation phenomena in earth–space telecommunications system design is illustrated in this section, and it appears desirable to include some related considerations about systems as well. The propagation loss L and system noise temperature T_{sys} appear in the link power budget equation, and reference to system design relates to link budgets.

The system designer may have the function of meeting system requirements posed by the user, but in the process of attempting to do so it may develop that the requirements present problems and may need to be modified. The design of a complicated system such as a telecommunications system is largely an iterative process, starting with a preliminary design, rather than a true synthesis. The amount of readily available information dealing specifically with system design is limited, but a useful treatment of the subject has been provided in the past.

Some minimum signal-to-noise ratio is needed for satisfactory operation of a telecommunications system, and information must be available or a decision must be reached in some way as to what this value is. Because of the characteristics of the propagation medium, S/N tends to be a random variable and, as it is usually impractical to design a system so that S/N never drops below any particular level, a specification should normally be made of the permissible percentage of time for which S/N may be below the specified level. This specification defines the signal availability, namely, the percentage of time that a specified S/N ratio should be available. Alternatively, or additionally, a specification may be made concerning outage: for example, the mean outage duration, or the time until the next outage. In some cases the statistical nature of the phenomena affecting S/N may not be known, and it may not be possible to design a system to have a specified availability or outage characteristic. In such cases, one may nevertheless need to estimate the margins that should be provided for the phenomena under consideration as best one can. For example, a margin of so many decibels must be allotted in some cases to take account of ionospheric scintillation, even though a satisfactory statistical description of the scintillation may not be available.

5.7.1 DIGITAL SYSTEMS

For digital systems performance is generally measured in terms of bit error rate, and the BER is a function of the energy-per-bit to noise-power-density ratio E_0/N_0. The energy per bit E_b is related to signal power S by $E_b R = S$, where R is the information rate in bits per second. Therefore,

$$E_b/N_0 = S/(N_0 R). \qquad (5.51)$$

Also,

$$(E_b/N_0)(R/B) = S/N. \qquad (5.52)$$

Equation (5.52) shows that if bandwidth B equals bit rate R, $E_b/N_0 = S/N$. The ratio R/B depends on the type of modulation and coding used. For uncoded binary phase-shift modulation (BFSK) employing phase values of $0°$ and $180°$, B may be equal to R. For uncoded quadriphase modulation (QFSK) employing phase values of $0°$, $90°$, $180°$, and $270°$, the bandwidth B may be only half the bit rate, as for each phase there are two corresponding bits. Coding of digital transmission is used as a means of minimizing errors or to reduce the needed E_b/N_0 ratio and therefore the power S needed for a fixed BER. Coding involves adding redundant symbols to an information symbol sequence and requires additional bandwidth beyond that of the original uncoded signal. The ratio of the number of information-bearing symbols to the total number is known as the rate of the code and has values such as 3/4 or 2/3, with 1/3 usually being the minimum value used. The two principal types of error correcting codes are block codes and convolutional codes. Forward error correction codes have applications for ameliorating the effect of attenuation due to rain. When coding is used in this way, a small amount of system capacity may be held in reserve and allocated as needed for links experiencing attenuation. The link data rate remains constant when following this procedure, the additional capacity being used for coding or additional coding. Although block codes may be used in some cases, convolutional codes have the advantages for satellite communications of ease of implementation and availability of attractive coding schemes. Convolutional coding and Viterbi decoding are an effective combination. The performance of a Viterbi decoder depends upon the rate R, the number K of consecutive information bits encoded (e.g., 4, 6, or 8), the levels of quantization Q (1 to 8), and path length (e.g., 8, 16, 32 bits). Figure 5.8 shows illustrative plots of BER vs E_b/N_0 for convolutional coding and Viterbi decoding and for no coding.

5.7.2 ANALOG SYSTEMS

The allowable noise in analog systems used for voice communications may be specified in pW0p, standing for noise power in picowatts at a point of zero

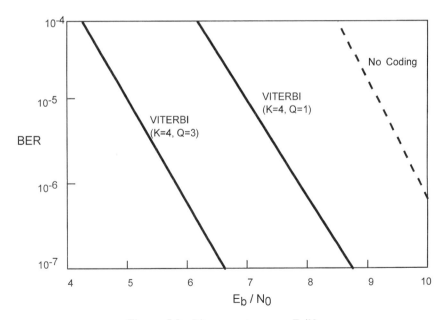

Figure 5.8 Bit error rate versus E_b/N_0.

relative level (0) with psophometric weighting (p) utilized. We consider here how the system designer, given the permissible value of pW0p, can determine the corresponding S/N ratio.

In recommendation 353-5, the CCIR advises that the noise power at a point of zero relative level in any telephone channel used in FDM-FM (frequency division multiplex–frequency modulation) telephony in the fixed satellite service should not exceed the following values:

10,000 pW0p for psophometrically weighted one-minute mean power for more than 20% of any month

50,000 pW0p for psophometrically weighted one-minute mean power for more than 0.3 percent of any month

1,000,000 pW0 for unweighted power (with an integration time of 5 ms) for more than 0.01 percent of any year

For RF levels above the FM threshold (commonly 10 dB above the noise level), the noise expressed in pW0p can be related to carrier power by

$$10 \log \text{pW0p} = -S_{\text{dBm}} - 48.6 + F_{\text{dB}} - 20 \log (\Delta f / f_{\text{ch}}), \qquad (5.53)$$

where F is the receiver noise figure, Δf is the peak frequency deviation of the channel for a signal of 0 test tone level, and f_{ch} is the center frequency occupied by the channel in the baseband.

Solving for S_{dBm} and then subtracting $10 \log KT = 10 \log KT_0 + F_{\text{dB}}$ $= 174 \text{ dBm} + F_{\text{dB}}$ for $T_0 = 290$ K yields

$$(S/N_0)_{\text{dB}} = (S/KT)_{\text{dB}} = 125.4 - 20 \log (\Delta f / f_{\text{ch}}) - 10 \log \text{pW0p}. \qquad (5.54)$$

5.8 Allocation of Noise and Signal-to-Noise Ratio

A communication satellite system consisting of an uplink and a downlink is subject to thermal noise in the uplink and downlink, to intermodulation noise generated in the satellite transponder in a system, and to interfering signals which may be received on the uplink or downlink or both. Considering all the individual noise sources to be additive at the downlink receiver input terminal, the ratio of signal power S to total noise power density $(N_0)_T$ is given by

$$S/(N_0)_T = \frac{S}{(N_0)_U + (N_0)_D + (N_0)_{\text{IM}} + (N_0)_I}, \qquad (5.55)$$

where $(N_0)_D$ is generated in the downlink, $(N_0)_{IM}$ represents intermodulation noise, and $(N_0)_I$ represents interference. The quantity $(N_0)_U$ is derived from but is not equal to the noise $(S_0')_U$ at the satellite (uplink) receiver input terminal. The relation between the two quantities is $(S_0)_U = (S_0')_U \, G/L_t$, where G is gain of the satellite transponder and L_t is the total downlink loss factor. It can be shown by some manipulations that

$$\frac{1}{(S/N_0)_T} = \frac{1}{(S/N_0)_U} + \frac{1}{(S/N_0)_D} + \frac{1}{(S/N_0)_{IM}} + \frac{1}{(S/N_0)_I}. \qquad (5.56)$$

The ratio $(S/N_0)_T$ appears at the downlink receiver input terminal; the ratio $(S/N_0)_D$ would be observed at this point if the input signal for the downlink was noiseless and interference was negligible. If one knows the values of the term of Equation (5.56) but one, that unknown quantity can be determined from (5.56). Noise allotted to the space segments includes noise generated in the uplink and downlink, intermodulation noise generated in the satellite transponder, and interference other than terrestrial interference.

5.9   Diversity Reception

Diversity reception of several types, most prominently site diversity, space diversity, and frequency diversity, may be advantageous for particular applications. For satellite communications site diversity can be used to reduce the effect of attenuation due to rain. Site diversity takes advantage of the fact that high rain rates tend to occur only over areas of limited extent. For example, the probability that rates greater than 50 mm/h will occur jointly at two locations 20 km apart is reported to be about 1/15th the probability that the rate will occur at one location. Most interest in site diversity is directed to frequencies above 10 GHz for which attenuation due to rain is most severe. For terrestrial line-of-sight paths, space and frequency diversity are used to combat fading due to atmospheric multipath and reflections from surfaces. The form of space diversity most used involves vertical separation of two receiving antennas on the same tower.

The performance of a diversity system can be characterized by diversity gain and diversity advantage, which are shown in Figure 5.9. Diversity gain is the difference, for the same percentage of time, between the attenuation exceeded on a single path and that exceeded jointly on two paths to two sites. Diversity advantage is defined as the ratio of the percentage of time that a given attenuation is exceeded on a single path to the percentage of time that a given attenuation is exceeded jointly on two paths. Site diversity to minimize the effects of attenua-

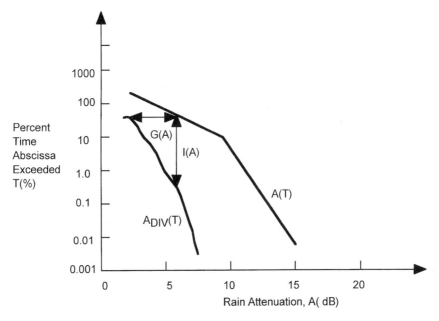

Figure 5.9 Defining diversity gain $G(A)$ and diversity advantage $I(A)$.

tion due to rain may be useful for critical applications at frequencies below 10 GHz, but must be weighted against the alternative of providing a margin to cover the extended attenuation. For higher attenuations that tend to occur at higher frequencies, the advantage is more apt to be on the side of site diversity.

The link power budget equation gives the received signal-to-noise ratio in terms of all the various parameters that affect it. Two of these, loss factor L and system noise temperature T_{sys}, tend to be random variables. The object of system is to ensure a satisfactory signal-to-noise ratio for a specified high percentage of time. The equation can be written in terms of $S/N_0 = (S/KT_{sys})$, where S is the signal power (W) and K is Boltzmann's constant. The ratio S/N_0 is the signal power to noise density ratio, as N_0 is the power per hertz. To obtain S/N from S/N_0, one can simply divide by B, the bandwidth in hertz. The quantity S/N_0 is written in the form

$$\frac{S}{N_0} = \frac{S}{KT_{sys}} = \frac{\text{EIRP } G_R}{L\, L_{FS}\, KT_{sys}}, \tag{5.57}$$

where EIRP stands for effective isotropic radiated power, G_R is the gain of the receiving antenna, L is a loss factor greater than unity if truly representing a loss, and L_{FS} is the free space basic transmission loss. The propagation medium plays a major role in determining L and T_{sys}. In carrying out satellite telecommunications system design, attention must be given to both the uplink and the downlink; both affect the S/N_0 ratio observed at the downlink receiver input terminal. For applying Equation (5.57), we separate EIRP into the product of P_T and G_T and convert to decibel values as is customary, with the result that

$$(S/N_0)_{dB} = (P_T)_{dB} + (G_T)_{dB} + (G_R)_{dB} - L_{dB} - (L_{FS})_{dB}$$
$$- K_{dBW} - (T_{sys})_{dB}, \tag{5.58}$$

where for K we actually use Boltzmann's constant K times 1 K times 1 Hz to obtain a quantity in dBW. Then T_{sys} and bandwidth B, when it is utilized, are treated as nondimensional. But G_R and T_{sys} are often combined into one term which is considered a figure of merit. Using this combination and also reverting back to EIRP,

$$(S/N_0)_{dB} = (EIRP)_{dB} + (G_R/T_{sys})_{dB} - L_{dB} - (L_{FS})_{dB} - K_{dBW}. \tag{5.59}$$

The system noise temperature T_{sys} is a measure of noise power, as N_0, the noise power per hertz, equals KT_{sys}, where K is Boltzmann's constant (1.38×10^{-23} J/K). Also, N, the total noise power, equals S_0 times bandwidth B. System noise temperature is defined at the antenna terminal of a receiving system as shown in Figure 5.10, which shows an antenna having a noise temperature of T_A, a lossy transmission line at the standard temperature T_0 (290 K), and a receiver having a noise temperature of T_R.

For the receiving system of Figure 5.10, T_{sys} is given by

$$T_{sys} = T_A + (L_a - 1)T_0 + L_a T_R. \tag{5.60}$$

To illustrate the calculation of T_{sys}, let T_R equal 100 K and $T_A = 50$ K, and consider that the transmission line loss is 1 dB. In the expression for T_{sys},

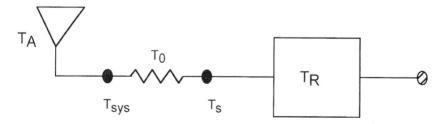

Figure 5.10 Receiving system showing location of T_{sys}.

$L_a = 1/G_a$, where G_a is less than unity and is the power "gain" of the transmission line, considering it as a lossy attenuator. The relation between G_a and attenuation A in decibels is

$$-A_{dB} = 10 \log G_a$$

and for $A = 1$,

$$-1 = 10 \log G_a, \quad G_a = 0.794, \quad L_a = 1/G_a = 1.26.$$

Substituting values into the expression for T_{sys},

$$T_{sys} = 50 + (1.26 \ 1) \ 290 + 1.26 \ (100) = 50 + 75.4 + 1.26 = 251.4 \text{ K}.$$

Note that if there were no attenuation between the antenna and receiver, G_a and L_a would equal unity and T_{sys} would equal $T_A + T_R = 150$ K.

The noise power density N_0 corresponding to $T_{sys} = 251.4$ K is given by

$$N_0 = K T_{sys} = (1.38 \times 10^{-23})(251.4) = 3.47 \times 10^{-21} \text{ W}$$

and

$$(N_0)_{dB} = 10 \log (3.47 \times 10^{-21}) = -204.6 \text{ dBW}.$$

Also,

$$(N_0)_{dBm} = -174.6 \text{ dBm}.$$

The quantity N_0 is 204.6 dB below 1 watt (W) and 174.6 dB below 1 milliwatt (mW).

The gain of an antenna having an effective aperture area A_{eff} is given by

$$G = (4 \pi A_{eff}) / \lambda^2,$$

where λ is wavelength. The effective area for the antenna aperture considered here equals the geometric area times an efficiency factor κ which generally falls between about 0.5 and 0.7 or higher. To illustrate antenna having a diameter of 3 m, and an efficiency factor of 0.54. The wavelength $\lambda = 3.0 \times 10^8 / (3.0 \times 10^9) = 0.1$ m and

$$A_{eff} = (\pi d^2 / 4) \kappa = \pi (9/4)(0.54) = 3.817 \text{ m}^2.$$

Thus,

$$G_{dB} = 10 \log G = 10 \log [4 \pi (3.817) / 0.01] = 36.8.$$

Calculating the distance and elevation angle of GEO satellite is done as follows. Consider an earth receiving station at 65° N and a GEO satellite on the same meridian. The distance d between the station and the satellite is given by

$$d_2 = r_0^2 + (h + r_0)^2 - 2r_0 (h + r_0) \cos \theta',$$

where r_0 is the earth radius, h is the height of the satellite above the earth, and θ' is latitude (65° in this example). Thus,

$$\begin{aligned} d^2 &= (6378)^2 + (35,786 + 6378)^2 - 2(6378)(35,786 + 6378)^2 \cos 65° \\ &= 39,889.6 \text{ km.} \end{aligned}$$

To determine the elevation angle θ, the following expression can be solved for Ψ, which equals the elevation angle θ plus 90°.

$$\begin{aligned} (h + r_0)^2 &= d^2 + r_0^2 - 2r_0 d \cos \Psi \\ (42,164)^2 &= (39,889.6)^2 + (6378)^2 - 2(6378)(39,889.6) \cos \Psi, \end{aligned}$$

where $\Psi = 106.67°$. The angle $\theta = 106.67° - 90° = 16.67°$.

If the earth station were displaced by 10° from the longitude of the satellite, then in place of $\cos \theta' = \cos 65° = 0.4226$ one would use $\cos 65° \cos 10° = 0.4162$. The result would be that $d = 39,932$ km and $\theta = 16.24°$. If the difference in longitude were 20°, the distance would be 40,060 km and the elevation angle would be 15°.

It may be necessary in system design to provide for service over a given extensive geographical area rather than for only a particular earth station. The relation between the service or coverage area, A_{cov}, and system parameters, including S/N_0, is shown in Eq. (5.61) as an approximation, from which K and other numerical factors were eliminated. The antenna efficiency factor κ_{ant} is used and A_R is the effective area of the receiving antenna:

$$A_{cov}(S/N_0) \approx \frac{P_T A_R \kappa_{ant}}{K T_{sys} L}. \tag{5.61}$$

References

1. R. F. Harington, "Field Computation by Moment Method," McMillan, 1968.

2. H. Law, "Performance Concerns Guide the Design of Pager Antennas," *Microwaves & RF,* December 1995.

3. G. Rosol, "Planning an Antenna Installation for a Wireless System," *Microwaves & RF,* October 1995.

4. J. M. Flock, "Method Determines the Performance of Dipole Arrays," *Microwaves & RF,* September 1995.

5. G. J. Stern and R. S. Elliot, "The design of microstrip including mutual coupling, Part II; Experimed," *IEEE Trans. on Antennas and Propagation,* Vol. 38, February 1990, pp. 145–151.

6a. J. A. Dooley, "New Passive Repeater Technology Fits Obstructed Areas for Cell-based System," *Wireless Systems Design,* Sept. 1996.

6b. B. Daly, "Cut Signal Losses for In-Building Systems," *Microwaves & RF,* August 1995.

7. W. C. Jakes, ed., "Microwave Mobile Communications," J. Wiley & Sons, 1974.

8. K. Feher, "Wireless Digital Communication," Prentice Hall, 1995.

9. P. S. K. Levny and K. Feher, "A Novel Scheme to Aid Coherent Dectection of GMSK Signals in Fast Rayleigh Fading Channels," *Proc. of the Second Int. Moble Sal, Conf.,* Ottawa, Canada, June 17–20, 1990, pp. 605–611.

10. CCIR Report 322, "World Distribution and Characteristics of Atmosphere Radio Noise," Published separately by the International Telecommunications Union, Geneva, 1964.

11. A. Boischot, "Les Bruits Cosmiques," AGARD Conference on Electromagnetic Noise, Interference and Compatibility, NATO, November 1975.

12. CCIR Report 670, "Worldwide Minimum External Noise Levels, 0.1 Hz to 100 Ghz," Green Book, Vol. 1, pp. 442–427, 1978. See also A. D. Spaulding and G. H. Hagn, "Worldwide Minimum Enironmental Radio Noise Levels (0.1 Hz to 100 Ghz)," pp. 177–182 of Symposium Proceeding: Effects of the Ionosphere on Space and Terrestrial System, J. M. Goodwin, ed. Arlington, Va., ONR/NRL, Jan. 24–26, 1978.

13. O. R. White, ed., "The Solar Output and Its Variation," Colorado Associated Universities Press, 1977.

14. R. F. Donnelly, ed., "Solar-Terrestrial Prediction Group Reports," Space Environment Laboratory, NOAA, Boulder, Co., August, 1979.

15. S. F. Smerd, "Non-thermal Sources and Amplified Emission in Solar Radio Bursts," in *The Solar Spectrum,* pp. 398–417. Edited by C. deJager, D. Reidel Publishing Co, Dordrecht-Holland, 1965.

16. A. Kruger, "Introduction to Solar Radio Astronomy and Radio Physics," *Geophys. Astrophys.* Mon. 16, Reidel, Dordrecht, 1979.

17. A. Maxwell, "The Variable Solar Radio Spectrum," in *The Solar Spectrum,* pp. 342–397. Edited by C. de Jager, D. Reidel Publishing Co., Dordrecht- Holland, 1965.

18. J. D. Kraus, "Radio Astronomy," McGraw-Hill, 1966.

19. K. W. Yates and R. Wielebinski, "Intensity Frequency Dependence of the Radio Sky Background," *Austr. J. Phys.* **19,** 389–407, 1966.

20. H. C. Ko and J. D. Kraus, "A Radio Map of the Sky at 1.2 Meters," *Sky and Telescope* **16,** pp. 160–161, 1957.

21. R. E. Taylor, "136 Mhz/400 Mhz Radio-Sky Maps," *Proc. IEEE* **61,** pp. 469–472, Apr. 1973.

22. ITT, Reference Data for Radio Engineers, Sixth Edition (1975), Chapter 29 Radio Noise and Interference, Howard W. Sams, Fourth Printing, 1981.

23. D. L. Jauncey, ed., Radio Astronomy and Cosmology, International Astronomical Union Symposium No. 74, Held at Cavendish Laboratory, August 16–20, 1976, D. Reidel Publishing Co., Dordrecht-Holland, 1977.

24. G. T. Haslum, W. E. Wilson, D. A. Grahan, and G. C. Hunt, "A Further 408 Mhz Survey of the Northern Sky," *Astron. Astrophys. Suppl.* **13,** pp. 359–394, 1974.

25. E. M. Berkhyijsen, "A Survey of Continuum Radiation at 820 Mhz Between Declinations 7° and + 85°." *Astron. Astrophys. Suppl.* **5,** pp. 263–312, 1972.

26. CCIR Report 720-1, "Radio Emission Form Natural Sources Above About 50 Mhz," Green Book, Vol. V, currently available as CCIR Doc. 5/5004, 5 Sept. 1981 and 5/1028 (XV Plenary Assembly, Geneva, 1982).

27. A. A. Penzias and R. W. Wilson, "A Measurement of Excess Antenna Temperature at 4080 Mc/S," *Astrophys. J.* **142,** 419–421, July 1965.

28. R. Wielebinski, "Antenna Calibration," Chapter 1.5 (pp. 82–97) of "Methods of Experimental Physics," Volume 12, ASTROPHYSICS, Part B Radio Telescopes, edited by M. L. Meeks, Academic Press, 1976.

29. W. E. Howard III and S. P. Maran, "General Catalogue of Discrete Radio Sources." *Astrophy J. Suppl.* Series No. 93, Vol X, p. 1, 1965.

30. R. G. Reeves et al. Manual of Remote Sensing (2 Volumes, R. C. Reeves, Editor-in-chief). American Society of Photogrammetry, Falls Church, VA, 1975.

31. J. W. Waters, "Emission and Absorption of Atmospheric Gases," Chapter 2.3 (pp. 142–176) of "Methods of Experimental Physics," Vol. 12, ASTROPHYSICS, Part B, Radio Telescopes, edited by M. L. Meeks, Academic Press, 1976.

32. E. K. Smith and J. W. Waters, "Attenuation and Sky Noise Temperature Due to the Gaseous Atmosphere Using the JPL Radiative Transfer Program—A Comparison of JPL and CCIR Values," JPL Publication 81, Jet Propulsion Laboratory, California Institute of Technology, Pasadena, CA, August 15, 1981.

33. L. W. Carrier, G. A. Cato, and K. J. von Essen, "The Backscattering and Extinction of Visible and Infrared Radiation by Selected Major Cloud Models," *Applied Optics* 6, p. 1209, July 1967.

34. S. D. Slobin, "Microwave Attenuation and Noise Temperature of Clouds at Frequencies Below 50 Ghz," Jet Propulsion Laboratory, California Institute of Technology, Pasadena, CA, July 1, 1981.

35. W. L. Flock, "Electromagnetics and the Environment: Remote Sensing and Telecommunications," Prentice Hall, 1979.

36. CCIR Report 721, "Attenuation and Scattering by Precipitation and Other Atmospheric Particles," Green Book, Vol. V, pp. 107–115, 1978. Currently available as CCIR Doc. 5/1029 of the XVth Plenary Assembly, Geneva, 1982.

37. L. J. Ippolitio, R. D. Kaul, and R. G. Wallace, "Propagation Effects Handbook for Satellite Systems Design: A Summary of Propagation Impairments on 10 to 100 Ghz Satellite Links with Techniques for System Design" (second edition), NASA Reference Publication 1082. National Aeronautics and Space Administration, Scientific Technical and Information Branch, 1981.

38. M. P. M. Hall, "Effects of the Troposphere on Radio Communication," Peter Peregrinus Ltd. on Behalf of the Institution of Electrical Engineers, 1979.

39. L. Tsand, J. A. Long, E. Njoku, D. H. Stealin, and J. W. Waters, "Theory of Microwave Thermal Emission From a Layer of Cloud or Rain," *IEEE Trans. Ant. and Prop.,* **Vol. AP–25,** pp. 650–657, September, 1977.

40. A. M. Zaody, "Effect of Scattering by Rain on Radiometer Measurements at Millimeter Wavelengths," *Proc. IEEE* **121,** pp. 257–263, April, 1974.

41. D. L. Carpenter, "Whistler Evidence of the 'Knee' in the Magnetospheric Ionization Density Profile," *J. Geophys. Res.,* **68,** 1675–1682, 1963.

42. C. R. Chappell, C. R. Baugher, and J. L. Horwitz, "The New World of Thermal Plasma Research," *Rev. Geophys. Space Phys.,* **18,** pp. 853–861, 1980.

43. C. R. Chappell, K. K. Harris, and G. W. Sharp, "A Study of the Influence of Magnetic Activity on the Location of the Plasmapause as Measured by OGO 5, 3." *Geophys. Res.,* **75,** pp. 50–56, 1970.

44. C. R. Chappell, K. K. Harris, and G. W. Sharp, "The Morphology of the Bulge Region of the Plasmasphere," *J. Geophys. Res.,* **75,** 3848–3861, 1970.

45. C. R. Chappell, K. K. Harris, and G. W. Sharp, "The Dayside of the Plasmasphere," *J. Geophys. Res.,* **76,** pp. 7632–7647, 1971.

46. K. I. Gringauz, "The Structure of the Ionized Gas Envelope of the Earth from Direct Measurements in the USSR of Local Charge Particle Concentrations," *Planet. Space Sci.,* **11,** pp. 281–296, 1963.

47. K. I. Gringauz, V. G. Kurt, I. Moroz, and I. I. Shklovskii, "Results of Observations of Charged Particles Observed Out to R = 100,000 km, with the Aid of Charged Particle Traps on Soviet Space Rockets," *Soviet Astronomy-AJ,* **4,** pp. 680–695, 1960.

48. D. L. Carpenter and C. G. Park, "On What Ionospheric Workers Should Know about the Plasmapause-Plasmasphere," *Rev. Geophys. Space Phys.,* **11,** pp. 133–154, 1973.

49. P. Decreau, "Fonctionnement d'une sonde equadripolaire sur satellite magnetospherique (experience GEOS). Contribution a l'etude du comportement du plasma froid au voisinage de la plasmapause equatoriale." These se Doctorat d'Etat en Sciences Physiques, Universite d'Orleans, 1983.

50. P. M. E. Decreau, C. Beghin, and M. Parrot, "Global Characteristics of the Cold Plasma in the Equatorial Plasmapause Region as Deduced from the GEOS 1 Mutual Impedance Probe, 3." *Geophys. Res.,* **87,** pp. 695–712, 1982.

51. C. R. Chappell, "Detached Plasma Regions in the Magnetosphere," *J. Geophys. Res.* **79,** pp. 1861–1870, 1974.

52. D. B. Muldrew, "F-Layer Ionization Troughe Deduced from Alouette Data," *J. Geophy. Res.,* **70,** pp. 2635–2650, 1965.

53. H. A. Talor, Jr., H. C. Brinition, D. L. Carpenter, F. M. Bonner, and R. L. Heyborne, "Ion Depletion in the High-latitude Exosphere: Simulations OGO 2 Observations of the Light Ion Trough and the VLF Cutoff," *J. Geophys. Res.,* **74,** pp. 3517–3528, 1969.

54. R. J. Hoch, "Stable Auroral Red Arcs," *Rev. Gep[hys. Space Phys.],* **11,** pp. 935–949, 1973.

55. J. E. Titheridge, "Plasmapause Effects in the Topside Ionosphere," *J. Geophys. Res.* **81,** pp. 3227–3233, 1976.

56. C. R. Chappell, J. L. Green, J. F. E. Johnson, and J. H. Waite, Jr., "Pitch Angle Variations in Magnetospheric Thermal Plasma–Initial Observations from Dynamics Explorer 1," *Geophys. Res. Letters,* **9,** pp. 933–936, 1982.

57. J. Lemaire, "The Mechanism of Formation of the Plasmapause," *Ann. Geophys.,* **31,** pp. 175–189, 1975.

58. J. Lemaire, Frontiers of the plasmasphere, These d' Agregation de l'Enseignement Superieur, UCL, Ed. Cabay, Louvain-La-Neuve, Aeronomics Acts An'298, 1985.

59. A. Papoulis, "The Fourier Integral and Its Applications," McGraw-Hill, New York, 1962.

60. C. R. Paul and D. D. Weiner, "A summary of required input parameters for emitted models in IEMCAP," RADC-TR-78-140.

61. Intrasystem Electromagnetic Compatibility Analysis Program (IEMCAP) F-15 Validation—Validation and Sensitivity Study; RADC-TR-77-290, Part I (of two), Final Technical Report, September 1977.

62. Intrasystem Electromagnetic Compatibility Analysis Program, Vol. I—User's Manual Engineering Section; RADC-TR-74-342, Final Report, December 1974.

63. S. A. Goldman, "Frequency Analysis, Modulation and Noise," McGraw-Hill, New York, 1948, pp. 76–79.

64. D. D. Weiner, "Discrete System Equations," IEMCAP Course Notes, RADC/RBCTI, Griffis AFB, New York, 1978.

65. "Handbook of Mathematical Functions with Formulas, Graphs and Mathematical Tables," U.S. Department of Commerce, NBS Applied Mathematical Series 55, p. 297, June 1964.

66. S. Skin and J. J. Jones, "Modern Communications Principles—with Application to Digital Signaling," McGraw-Hill, New York, 1967, p. 139.

67. Thomas E. Baldwin et al. "Intrasystem Analysis Program Model Improvement," RADC-TR-82-20, Rome Air Development Center, Griffiss AFB, NY, Dec. 1981.

68. Interference Notebook, RADC-TR-66-1, Contract AF 30(602)-3118; AD 484-485, June 1966.

69. "Adjacent Channel Interference Analysis," RADC-TR-68-595, Contract F30602-68-C-0061, March 1969.

70. J. F. Spina and D. D. Weiner, "Sinusoidal Analysis and Modeling of Weakly Nonlinear Circuits," Van Nostrand Reinhold, New York, 1980.

71. "Voice Communication Degradation Studies," Jansky & Bailey Engineering Dept., a Division of Atlantic Research Corp., RADC-TR-67-546, Contract No. AF 30(602)-4023.

72. V. Volterra, "Theory of Functional and of Integral and Integro-differential Equations," Blackie and Sons, London, 1930.

73. N. Wiener, "Response of a Non-linear Device to Noise," M.I.T. Radiation Laboratory Report V-165, April 1942.

74. Nonlinear System Modeling and Analysis with Applications to Communication Receiver, RADC-TR-73-178, June 1978.

75. W. C. Duff, "EMC Figure of Merit for Receivers," 1969 IEEE EMC Symposium Proceeding, Asbury Park, NJ, 1969.

76. 3rd Order Intermodulation Study, RADC-TR-67-344, ITT Research Institute, July 1967.

77. C. A. Wass, "A Table of Intermodulation Products," *Journal of the Institute of Electrical Engineering (London),* Part III, 1948.

78. G. T. Caprano, "Spurious Response Identification and Generation," Technical Memorandum No. EMC-TM-67-15, Rome Air Development Center, December 1967.

79. "Interference Analysis Study," RADC-TR-63-27, June 1963.

80. G. J. Myers, "Software Reliability," Wiley Interscience, New York, 1976.

81. J. H. McMahon, "Interference and propagation formulas and tables used in the FCC Spectrum Management Task Force Land Mobile Frequency Assignment Model," *IEEE Trans. Vehicular Technol.,* November 1974.

82. L. J. Ippolito, R. D. Kaul, R. G. Wallace, "Propagation Effects Handbook for Satellite System Design," 2nd ed. NASA Reference Publication 1082, Dec. 1981.

83. R. K. Crane, "A review of transhorizon propagation phenomena," *Radio Sci.,* **16,** 649–669, Sept.–Oct. 1981.

84. H. T. Dougherty, A Consolidated Model for UHF/SHF Telecommunication Links Between Earth and Synchronous Satellite, NTIA Report 80-45, U.S. Dept. of Commerce, August 1980.

85. CCIR, "The evaluation of propagation factors in interference problems between stations on the surface of the earth at frequencies above about 0.5 GHz," Report 569-3, Vol. V, Propagation in Non-ionized Media, Recommendations and Reports of the CCIR. Int. Telecomm. Union, Geneva, 1986.

86. CCIR, "Propagation data required for the evaluation of coordination distance in the frequency range 1 to 40 GHz," Report 724-2, Vol. V, Propagation in Non-ionized Media, Recommendations and Reports of the CCIR. Int. Telecomm. Union, Geneva, 1986.

87. CCIR, "Determination of coordination area," Report 382-5, Vols. IV and IX—Part 2, Fixed Service Using Radio Relay Systems. Frequency Sharing and Coordination between Systems in the Fixed Satellite Service and Radio-relay Systems. Recommendations and Reports of the CCIR. Int. Telecomm. Union, Geneva, 1986.

88. CCIR, "Propagation data required for evaluating interference between stations in space and those on the surface of the earth," Report 885-1, Vol. V, Propagation in Non-ionized Media, Recommendations and Reports of the CCIR. Int. Telecomm. Union, Geneva, 1986.

89. H. T. Dougherty and B. A. Hart, "Recent progress in duct propagation predictions," *IEEE Trans. Antenna Propagat.* **AP-27,** 542–548, July 1979.

90. L. J. Battan, "Radar Observations of the Atmosphere," Chicago University Press, 1973.

91. CCIR, "Propagation required for the evaluation of coordination distance in the frequency range 1 to 40 GHz," Report 724-1, Vol. V, Propagation in Non-ionized Media, Recommendations and Reports of the **CCIR.** Int. Telecomm. Union, Geneva, 1982.

Index

Note: Page numbers in italics refer to the figure or table on that page.